Hazardous Waste Management and Health Risks

Edited by

Gabriella Marfe

University of Campania "Luigi Vanvitelli",
via Vivaldi 43,
Caserta 81100,
Italy

&

Carla Di Stefano

Department of Hematology, "Tor Vergata" University, Viale
Oxford 81,
00133 Rome,
Italy

GRAPHICAL ABSTRACT

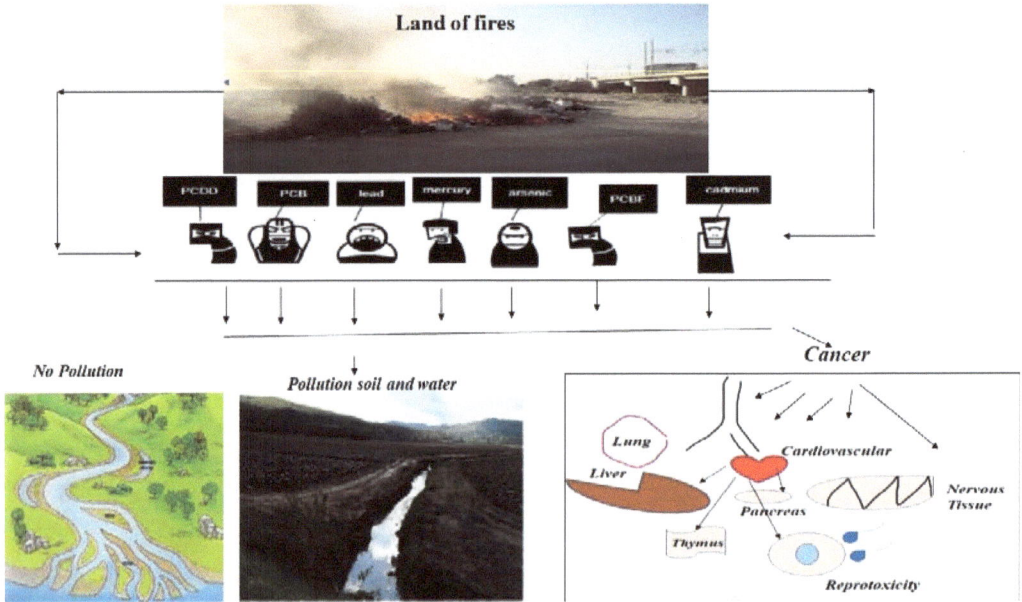

GRAPHICAL ABSTRACT

Hazardous Waste Management and Health Risks

Editors: Gabriella Marfe and Carla Di Stefano

ISBN (Online): 978-981-14-5474-5

ISBN (Print): 978-981-14-5472-1

ISBN (Paperback): 978-981-14-5473-8

need for a court order if at any point you breach any terms of this License Agreement. In no event will any delay or failure by Bentham Science Publishers in enforcing your compliance with this License Agreement constitute a waiver of any of its rights.

3. You acknowledge that you have read this License Agreement, and agree to be bound by its terms and conditions. To the extent that any other terms and conditions presented on any website of Bentham Science Publishers conflict with, or are inconsistent with, the terms and conditions set out in this License Agreement, you acknowledge that the terms and conditions set out in this License Agreement shall prevail.

Bentham Science Publishers Pte. Ltd.
80 Robinson Road #02-00
Singapore 068898
Singapore
Email: subscriptions@benthamscience.net

BENTHAM SCIENCE

CONTENTS

PREFACE i

FOREWORD ii
    REFERENCES vi

LIST OF CONTRIBUTORS ix

CHAPTER 1 INTRODUCTION 1
    Gabriella Marfe, Stefania Perna and *Carla Di Stefano*
    INTRODUCTION 1
    NOTES 10
    CONSENT FOR PUBLICATION 10
    CONFLICT OF INTEREST 10
    ACKNOWLEDGEMENTS 11
    REFERENCES 11

CHAPTER 2 HAZARDOUS WASTE AND ITS MANAGEMENT 17
    Gabriella Marfe, Stefania Perna1 and *Carla Di Stefano*
    INTRODUCTION 17
    HAZARDOUS WASTE 18
        Hazardous Waste Definition 18
        Exclusive List System 20
        Inclusive List System 20
        Characteristic Hazardous Waste 20
        Hazardous Waste Generation and Impacts 22
        Hazardous Waste Management 23
    LANDFILL DISPOSAL OF HAZARDOUS WASTE 25
        Landfill Disposal 25
        Methods of Landfill Disposal 25
        Selection of the Site 26
        Problems with Landfill Disposal 29
    INCINERATION OF HAZARDOUS WASTE 30
        Methods of Incineration 30
        Site Selection Models 31
        Site Selection 33
    STATUS OF HAZARDOUS WASTE MANAGEMENT 33
    CONCLUSION 42
    CONSENT FOR PUBLICATION 43
    CONFLICT OF INTEREST 43
    ACKNOWLEDGEMENTS 43
    REFERENCES 43

CHAPTER 3 HAZARDOUS WASTE AND ITS ASSOCIATED HEALTH RISKS 51
    Rishi Rana and *Rajiv Ganguly*
    INTRODUCTION 51
    HAZARDOUS WASTE: IDENTIFICATION AND CLASSIFICATION 53
    HAZARDOUS WASTE: IMPACT ON HEALTH 55
    HAZARDOUS WASTE: MANAGEMENT AND DISPOSAL 57
    CONCLUSION 59
    CONSENT FOR PUBLICATION 59
    CONFLICT OF INTEREST 59
    ACKNOWLEDGEMENTS 59

REFERENCES 59

CHAPTER 4 BIOLOGICAL EFFECTS OF HAZARDOUS WASTE: THRESHOLD LIMITS OF ANOMALIES AND PROTECTIVE APPROACHES 62
Amal I. Hassan and *Hosam M. Saleh*
INTRODUCTION 62
THE THRESHOLD OF HAZARDOUS MATERIALS IN THE BIOLOGICAL SYSTEMS 64
CHEMICALS OF PUBLIC HEALTH CONCERN AND PROPERTIES OF HAZARDOUS WASTES 73
NUCLEAR OR RADIOACTIVE WASTE 75
RELATIONSHIP OF HAZARDOUS WASTES AND CANCER MORTALITY 76
REPRODUCTIVE EFFECTS 77
HEALTH AND SAFETY OF WORKERS 78
BIOLOGICAL IMPACTS ASSOCIATED WITH HAZARDOUS WASTES 81
GENETICS EFFECTS 83
CYTOTOXIC MECHANISMS OF HAZARDOUS WASTES 85
THE REMEDIATION OF UNDESIRABLE EFFECTS OF HAZARDOUS WASTES ON HUMAN HEALTH 86
GREEN CHEMISTRY AS A NEW TREND OF REMEDIATION 88
CONCLUSIONS 90
CONSENT FOR PUBLICATION 90
CONFLICT OF INTEREST 91
ACKNOWLEDGEMENTS 91
REFERENCES 91

CHAPTER 5 HEAVY METAL POLLUTION: SOURCES, EFFECTS, AND CONTROL METHODS 97
Sunil Jayant Kulkarni
INTRODUCTION 97
SOURCES OF HEAVY METALS 98
EFFECT ON HUMAN HEALTH 99
REMOVAL METHODS 101
CONCLUSION 103
CONSENT FOR PUBLICATION 104
CONFLICT OF INTEREST 104
ACKNOWLEDGEMENT 104
REFERENCES 104

CHAPTER 6 REPRODUCTIVE BIOMARKERS AS EARLY INDICATORS FOR ASSESSING ENVIRONMENTAL HEALTH RISK 113
Luigi Montano
INTRODUCTION 114
THE CHALLENGES OF CONTAMINATED SITES REMEDIATION 118
BIOMARKERS FOR THE EVALUATION OF HEALTH RISK 121
Biomarkers' Interpretation 122
Biomarkers' Benefits 123
Biomarkers' Classification in the Interaction Between Organism and Contaminant 124
Specificity of the Response 125
Biomarkers as a Diagnostic Tool 125
FROM BIOMARKERS TO SENTINEL ORGANS TO DETECT EARLIEST RISK INDEXES 126

THE IMPACT OF CONTAMINANTS ON THE MALE REPRODUCTIVE SYSTEM: THE
SENTINEL ORGANS 126
EPIGENETIC DAMAGES OF ENVIRONMENTAL CONTAMINANTS: THE SPERM
EPIGENOME 129
HUMAN SEMEN: ENVIRONMENTAL AND HEALTH MARKER 131
CONCLUSION 134
CONSENT FOR PUBLICATION 134
CONFLICT OF INTEREST 134
ACKNOWLEDGEMENTS 134
REFERENCES 135

CHAPTER 7 HAZARDOUS WASTE, HEALTH PROBLEMS, AND PERSONAL WELL-
BEING: AN INTERNATIONAL PERSPECTIVE 146
    James G. Linn, Debra R. Wilson, Jorge Chuaqui and Thabo T. Fako
    INTRODUCTION 146
    DEFINITION OF TERMS 147
        Hazardous Waste 147
        Health Problems Associated with Hazardous Waste 148
        Personal Well-being 148
        Hazardous Waste Crises, Sustainable Economic Development, and Environmental Health   148
            Sustainable Economic Development 148
        Environmental Health 149
        Hazardous Waste Crises & Radioactive Accidents in Industrialized and Developing
        Societies 150
        Hazardous Waste, Intellectual Development, Mental Illness and Subjective Well-being 152
        The Way Forward to Reduce the Impact of Hazardous Waste on Physical Health, Mental
        Health, Intellectual Development and Personal Wellbeingg 154
    CONCLUDING REMARKS 154
    CONSENT FOR PUBLICATION 155
    CONFLICT OF INTEREST 155
    ACKNOWLEDGEMENTS 155
    REFERENCES 155

CHAPTER 8 ZERO WASTE MANAGEMENT AS KEY FACTOR FOR SUSTAINABLE
DEVELOPMENT 159
    Gabriella Marfe, Carla Di Stefano and Arturo Hermann
    OBJECTIVE WITHIN THE GLOBAL INITIATIVES FOR SUSTAINABLE
    DEVELOPMENT 159
    THE UN AGENDA 2030 FOR SUSTAINABLE DEVELOPMENT 160
        The Historical Background 160
    MAIN FEATURES OF THE AGENDA 2030 161
        "Our Vision" 161
        "A Call for Action to Change our World" 161
    THE SUSTAINABLE DEVELOPMENT GOALS (SDGS) 162
    THE LINKS WITH ZERO WASTE INITIATIVES 165
    WASTE MANAGEMENT IN LIGHT OF THE SDGS 165
    SUSTAINABLE WASTE MANAGEMENT 171
    THE CONCEPT OF SUSTAINABLE WASTE MANAGEMENT 171
    ZERO WASTE AS THE IDEAL CONCEPT IN SUSTAINABLE WASTE
    MANAGEMENT 172
    THE CONCEPT OF ZERO WASTE 173

WASTE ELECTRIC AND ELECTRONIC EQUIPMENT (WEEE) IN ZERO WASTE STRATEGY 179

ZERO WASTE PRACTICES ACROSS THE WORLD 182

ZERO WASTE PRACTICES IN DEVELOPING COUNTRIES 183

CONCLUSIONS 185

CONSENT FOR PUBLICATION 186

CONFLICT OF INTEREST 186

ACKNOWLEDGEMENTS 186

REFERENCES 186

SUBJECT INDEX 193

PREFACE

Hazardous waste management is an important issue for environmental and public health in each country and it is a challenge both in the developed and developing countries. Today, this waste is poisoning the environment and causing damage to the health of both humans and animals. This book focuses on several aspects of hazardous waste, from its sources to its consequences, for human health and natural environment. Furthermore, this book explains how the release of chemical toxins derived by hazardous waste can be negatively implicated in environmental and health impacts. Using many examples and illustrations, it shows the relationship between ecological improper hazardous waste management (such as e-waste and hospital waste) and human health. The book starts with a brief introduction, presenting the management of hazardous waste and increasing volume of different research on health impact. It comprises of eight chapters where the third and fourth chapters describe the impact on human health associated with an inadequate hazardous waste management.

The fifth chapter describes the correlation between heavy metal pollution and various acute and chronic diseases in human beings, while sixth chapter illustrates how reproductive biomarkers can be considered as early signals of environmental pressure and increased risk of adverse chronic health effects. Then chapter seventh assesses the relationship between the observed health risks and environmental contamination from an international perspective. Finally, the last chapter illustrates briefly zero-waste strategy as possible alternatives in solving the problems of waste generation. The authors emphasize the need to consider the environmental impact of human activities and include some real-world examples.

Dr. Gabriella Marfe
Department of Scienze e Tecnologie Ambientali
Biologiche e Farmaceutiche, University of Campania "Luigi Vanvitelli,"
via Vivaldi 43, Caserta 81100
Italy

&

Dr. Carla Di Stefano
Department of Hematology
"Tor Vergata" University, Viale Oxford 81
00133 Rome
Italy

FOREWORD

Never has it been more necessary for such a book to investigate the effects of the increasingly relentless human creation of dangerous substances, endangering clean air, soil and sea, on which all life depends in a completely reckless and boundless way, severely damaging the habitats of all living things on earth and causing a 6th ever mass extinction of species. In fact, Paul Ehrlich, co-author of the "The Population Bomb Revisited [1] (quoted in Carrington D.P. 22, 03, 2018 In the Guardian Newspaper) [2] warned about the chemical pollution.

"The evidence we have is that toxics reduce the intelligence of children as members of the first heavily influenced generation are now adults." He treats this risk with characteristic dark humour: "The first empirical evidence we are dumbing down Homo sapiens", were the Republican debates in the US 2016 presidential elections – and the resultant '*kakistocracy*'.

On the other hand, toxification may solve the population problem, since sperm counts are plunging. A proper understanding of toxicology has become most urgent since the original realisation of the causes of Minamata disease in 1932 at the Chisso industrial plant, when people were contaminated in Japan by mercury poisoning from eating contaminated fish and sea food and became very ill as a result. The effects included numbness and severe neurological life changing problems [3a,b]. These risks have not gone away as mercury, amongst other substances, is still discharged into many rivers and seas. We do not seem to have learned how easy it is to affect human health by indiscriminate and irresponsible waste policies. Iconic disasters include Seveso in 1976, in Italy an explosion of tetrachlorodibenzo-p-dioxin (TCDD), a kind of dioxin owned by Roche. Carcinogenic effects of dioxins have been described and documented at high dose exposures, such as after the Seveso accident [4 - 16]. The disaster at the Union Carbide plant in Bhopal in India was terrible. The plant produced the carbamate carbaryl, also known as Sevin that was not particularly hazardous substance, but the chemical methyl-isocyanate (MIC) which was used in its production was extremely toxic. On December 2, 1984 a lethal cloud of Methyl Isocyanate (MIC) leaked from plant and its release covered approximately nine miles and consisted of 40 tons of MIC, causing an estimated 2500 immediate fatalities and more than 10,000 deaths within a week of the event [17, 18]. Furthermore, Love Canal (a neighborhood in Niagara Falls) was used to dump waste of World War II (1940) and then, in 1942, Hooker Chemical released an estimated 21,000 tons of hazardous chemical waste (exachlorocyclohexanes, benzylchlorides, organic sulfur compounds, chlorobenzenes, and sodium sulfide/sulfhydrates) causing many illnesses [19 - 23]. With regard to incineration plants, Medical Students are still not being taught about the health dangers, according to a report by Drs. Thompson and Antony [24], even though it is known that the area around Heathrow airport produces a danger with its toxic mix of aircraft fumes, car pollution from the busy M25 Motorway. and combined with the very large waste incinerator, it keeps the chronic obstructive pulmonary disease (COPD) ward extremely busy in the nearby hospital, to which I was kindly invited to speak and to meet people who were affected very badly by this toxic waste and mixing with other pollutants [25 - 28].

On the other hand, it is becoming increasingly apparent that Social Prescribing as it is now called, is the latest trend in health care, which is the idea that being part of and being able to experience. Nature is something which can actually heal the sick and enhances human health, including mental health, rather than destroying it. At the other end of the spectrum, some of the most hazardous waste products being so called 'thrown away', into the environment, harm animals, fish, marine life, and the climate itself and of course can be very damaging indeed to human health. Many species are going extinct very fast and before extinction species become weaker in many respects. Hazardous health risks contribute to this situation and we need to be most careful that it does not happen to us. In fact, *the hippocratic oath* advises Medical Practitionesr that they must always try to preserve life. Would it not be best if something similar was created for people at the moment when they are introducing the use of hazardous chemicals?

The issues with hazardous waste are many and varied but also include spatial movement of the waste from the original contamination site and can turn up anywhere around the globe. For example, hydrochlorofluorocarbons (HCFCs) and chlorofluorocarbons (CFCs) harm the ozone layer far above the earth, miles from the original site of their use. Temporally too, the modern chlorine family can last a long time in the atmosphere, and nuclear waste has the distinction that the period of its half-life, is known to cause damage to human health, actually lasts longer, in some cases, than the of the whole of human civilisation itself. The nuclear facility at Sellafield has been radioactive waste in the Irish Sea for years and this is now known to have caused and still be causing absolutely devastating and cruel life long debilitating and ghastly mutations and health effects borne by children on the receiving end of this [29 - 31]. We must surely be able to behave better than this and to sort out our waste in a more responsible way. Such hazards are very, very long lasting and the ethics of leaving such dangerous waste for thousands of human and other species' generations to come are very problematic and simply cannot be justified at all. Even the storage facilities are known to be vulnerable to seismic movements over time, or may not last at all and so are no guarantee of safety for future generations. If we don't want to be absolutely hated and vilified by future generations, then we need to sort out the issue of hazardous chemicals with great urgency.

The dangers of all forms of hazardous waste are very poorly understood and as with much of the economy, the attitude seems to pervade that if someone has paid for something or is making money out of a transaction, then it must by definition be 'a good thing for everyone in the community as a whole' This attitude has been justified, for nearly a century, by the morally and factually dubious 'trickledown theory.' For example, there was a huge fire last week, at the Lubrizol factory in Rouen, which residents say may have even polluted the River Seine, and which is owned by one of the world richest men. The local people felt the effects of the black soot pollution but not the owner of the factory, who typically often doesn't even live in the country where the pollution is being generated and who stands to make a great profit from the chemicals [32, 33]. Unfortunately, the effects actually go much wider for example, when one company decides to go fracking, other people in the community can find that the resultant waste products can come up through the water table and right into the taps in their houses, even causing fires. When a nuclear power plant decides to circulate cooling water or sending toxic waste into local pipelines, the local community more often than not, find there are health effects, which deal cope with and subsidize. As with many things in nature, everything circulates around the globe, someone can throw a plastic bag away in one part of the world and it can turn up in an ocean current on the other side of the world, and its effects are definitely not costed or planned.

The issue of Hazardous Waste and Health Risks is very dear to my heart having possibly suffered a devastating and late stage miscarriage possibly attributable to The Chernobyl Nuclear Plant and its hazardous waste. Today everyone on earth is affected by these issues. No animal or person is immune from pollution from the world's hazardous and human made chemicals and substances. We now know that soon there will be more plastic in the sea than fish, and that most sea birds are full of items such as cigarette lighters, boxes of plasters and all sorts of other plastic debris. Many whales are dying due to the ingestion of countless plastic bags. In our own bodies, our breast milk is full of all sorts of waste and effluent.

The issue is that so called 'waste' is 'flushed away' from the process in which it is being used. Whilst we are flushing 'away' something we regard as 'waste', we are also unreasonably expecting that our beautiful planet will somehow act as a global dustbin and 'get rid of it,' whilst, someone else on the other side of the world is 'throwing something else away' imagining then it will somehow spontaneously combust into a puff of smoke and disappear. However, the world is not like that, nothing vanishes, it just pops up somewhere else. Everything we do and everything we regard as waste goes somewhere. It goes somewhere else. And in so doing, it can cause unintended consequences and harm to other people, most of whom are not even involved in the use of that product or substance.

In this context, nuclear waste is now known to cause all sorts of cancers, some to the workers at the plant but also to the surrounding people or fauna. We also know that the precautionary principle is useful in helping prevent the risks that we cannot envisage. For example, when the steam engine was invented, no one ever expected that global warming would become one of the most pressing human health risks of our age. When plastic was invented, no one ever suspected the damage it would do to marine life. When DDT (dichlorodiphenyl-trichloroethane) was invented no one ever thought that spring would become silent and it would wipe out the birds and weaken their eggs, as reported by Rachel Carson in her important book, *Silent Spring* [34 - 38].

The point is that one person's utility maximization and use of a product or process, is another person's hazardous waste. We were unfortunately philosophically arrogant enough to believe that notion that we are the stewards of the earth and all of nature bows down to our species. We have found out too late, that this is not only the case but that nature fights back and that what we do to nature has a habit of coming back to haunt us too.

We have systematically destroyed the habitat of many of our fellow sister and brother creatures and never stopped to think that we are in fact just another creature. In some ways we are rather stupid. There is nothing in nature that says that the current mass extinction we have managed to cause ourselves will spare our species especially. In fact, the larger and more complex our society, actually the less resilient it becomes to issues, disasters and pollution.

Our so called civilisation is in fact relatively new, only 10,000 years old and by rocking its very foundations, reducing land mass, creating sea level rise and huge dead zones in the seas, we are reducing drastically our own habitat just at a time when our species has increased in number. Similarly by squandering many of the earth's wonderful resources, we have depleted the materials we depend on and chucked them away as 'waste', when they could much more easily be reused, recycled and repaired so that no more of earth's finite resources are plundered. As Boulding says, 'anyone who thinks that we can have infinite growth is either a madman or an economist'.

It seems people are beginning to understand that living lightly on the earth might be the only pathway to the future., For example when a hurricane, devastated Barbuda, the commissioner told me that instead of building brick houses, they would now be building them with grass,. Someone else told me that stones are now becoming the building bricks of choice. Cement has had a powerful health impact and we may well have to revert to earlier less dangerous building methods.

Ozone depleting chemicals are still causing holes in the ozone long after they have done the job they were intended to do. According to the Journal Nature, the latest real concern is the 4,730 (per- and polyfluoroalkyl substances)-PFAS-related structures from patent filings and chemical registries, ([https://www.atsdr.cdc.gov/toxprofiles/tp200.pdf), as well as a range of very serious health problems linked to PFOA (perfluorooctanoic acid) and PFOS (perfluorooctane sulfonic acid) [39 - 43]. The problem of waste is that we throw it away, and it goes to landfill and then the chemicals leach into the hydrology systems of all kinds, where they pollute and cause serious health effects either directly or indirectly.

Understanding the impacts of hazardous waste and applying the precautionary principle in all our work is one way to start addressing these problems. Also for humans to admit once and for all that we simply do not and never will know everything about nature and how it works, means a bit more humility when we find new wonder substances. Planning for unintended consequences might be a start to a more successful relationship with nature and the roulette which is human health. The solution needs to be to use all hazardous chemicals with great caution and not to use them to make money or products from the supply side, only to use them when we really need them and to start to consider alternatives or not using them at all. Many products we have become accustomed to, might after real consideration be found to be not worth those risks. I always mention that in Tudor times- around the year, 1600 ladies thought it was a good idea to whiten their faces with lead and they gave themselves lead poisoning. When people realised how dangerous it was, they stopped doing it. Today no one would be so silly as to do this. And so it is with hazardous chemicals, we live in an age where it is considered necessary to enjoy their so called benefits, but a new generation of young people is arising who are more frugal and more circumspect about unnecessary consumption We need to listen to them and start to end our love affair with dangerous waste and to question the use of hazardous materials: just because we might do it or decide not to do so.

As we now know, human created, anthropogenic pollution has a direct effect in the spread, severity and spatial layout of pandemics, a situation which is affecting the whole of humanity, and from which every single human is engaged in a battle for avoidance and survival. Our entire global economy ground to a halt for a while as a result in 2020. This really matters, this really affects everything. In particular, it is now likely, that air pollution can be mapped directly onto areas of case density, intensity and spread, using the vector of human health weakening and thus predisposition to vulnerability. This tends to happen in areas where humans have the least good health outcome predictions, due to socio economic factors, which is well documented in the literature and where humans are crammed together at the highest densities.

This has historical roots in the incoming populations, externally encouraged economic urbanisation, (from farming and the land) in order to provide mass labour for economics of scale in the industrialisation of whole economies to create large profits for those with investment capabilities. The process, which is now several 100s of years old, has reached the situation of creating the perfect storm. Only by understanding the role of pollution of all kinds in weakening the bodies and minds, (pollution is now even possibly implicated in dementia), of every single human on the planet, and every other living species can we hope to survive at all. Many of these other of our sister earth dwelling species are also today experiencing similar global diseases, because of our own pollution creating activities, (not theirs), from the bees, to trees and to bananas, can we hope to prevent the recurrence and spread of disease and for life on earth to survive. Apart from the deep moral questionability of this situation, from a purely selfish point of view, our own individual choices really do matter and giving populations access to the right information to make the right choices is now fundamental to survival. This book is a really vital piece of the jigsaw in providing a framework for the collection, debate and dissemination of the issues of absolutely critical information in an age where there is, of course, a deliberate attempt to stop the public having access to the truth and agency to change consumption and behaviour patterns.

I recommend this book to you and hope you enjoy the ideas and new perspectives that it raises and we all hope I am sure that it will contribute to bringing the human species to some kind of realisation that it has a responsibility to all of nature and to its own kind, to clean up its act and stop polluting our wonderful clean air, soil and water and to use nature more maturely and more beneficially for ourselves, our own health and for all of nature.

REFERENCES

[1]   Ehrlich PR, Ehrlich AH. The Population Bomb Revisited. Elect J Sustain Develop 2009; 1(3). Accessed at: https://www.populationmedia.org/wp-content/uploads/2009/07/Population-Bomb-Revisited-Paul-Ehrlich-20096.pdf 28/07/2020.

[2]   Carrington D. Interview with Paul Ehrlich: Collapse of civilization is a near certainty within decades. Fifty years after the publication of his controversial book The Population Bomb, biologist Paul Ehrlich warns overpopulation and overconsumption are driving us over the edge. Guardian's Overstretched Cities Series. D.P. 22, 03, 2018; In the Guardian on line: https://www.theguardian.com/cities/2018/mar/22/collapse-civilisation-near-certain-dec-des-population-bomb-paul- ehrlich 22.3.2018. Accessed 28/07/2020.

[3a] Hachiya N. The history and the present of minamata disease. Entering the second half a century. Japan Med Assoc J 2006; 49(3): p.112-8. The National Institute for Minamata Disease.

[3b] Hirofumi A., Policy implications toward Green Economics in Pollution prevention: theory and problems in Japan, in International Journal of Green Economics, Inderscience Publishers, Geneva 2006: Vol. 1(1), p174-200.

[4]   Warner M, Eskenazi B, Mocarelli P, *et al.* Serum dioxin concentrations and breast cancer risk in the Seveso women's health study. Environ. Health Perspect. 2002; 110, Warner M., Eskenazi B., Mocarelli P., *et al.* Serum dioxin concentrations and breast cancer risk in the Seveso Women's Health Study. Environ Health Perspect. 2002 Jul; 110(7): 625–628. Research Study. Environ Health Perspectives v.110(7); 2002 Jul PMC1240906 https://www.ncbi.nlm.nih.gov/pmc/articles/PMC1240906/. Accessed 28/07/2020.

[5]   Consonni D, Pesatori AC, Zocchetti C, *et al.* Mortality in a population exposed to dioxin after the Seveso, Italy, accident in 1976: 25 years of follow-up. Am J Epidemiol 2008; 167, 847–858.

[6]   Pesatori AC, Consonni D, Rubagotti M, Grillo P, Bertazzi PA. Cancer incidence in the population exposed to dioxin after the "Seveso accident": twenty years of follow-up. Environ Health 2009; 8: 39. [http://dx.doi.org/10.1186/1476-069X-8-39] [PMID: 19754930]

[7]   Warner M, Mocarelli P, Samuels S, Needham L, Brambilla P, Eskenazi B. Dioxin exposure and cancer risk in the Seveso Women's Health Study. Environ Health Perspect 2011; 119(12): 1700-5. [http://dx.doi.org/10.1289/ehp.1103720] [PMID: 21810551]

[8]   Ye M, Warner M, Mocarelli P, Brambilla P, Eskenazi B. Prenatal exposure to TCDD and atopic conditions in the Seveso second generation: a prospective cohort study. Environ Health 2018; 17(1): 22. [http://dx.doi.org/10.1186/s12940-018-0365-2] [PMID: 29482571]

[9]   Eskenazi B, Warner M, Brambilla P, Signorini S, Ames J, Mocarelli P. The Seveso accident: A look at 40 years of health research and beyond. Environ Int 2018; 121(Pt 1): 71-84. [http://dx.doi.org/10.1016/j.envint.2018.08.051] [PMID: 30179766]

[10] Eskenazi B, Mocarelli P, Warner M, *et al.* Seveso Women's Health Study: a study of the effects of 2,3,7,8-tetrachlorodibenzo-p-dioxin on reproductive health. Chemosphere 2000; 40(9-11): 1247-53. [http://dx.doi.org/10.1016/S0045-6535(99)00376-8] [PMID: 10739069]

[11] Eskenazi B, Mocarelli P, Warner M, *et al.* Seveso Women's Health Study: does zone of residence predict individual TCDD exposure? Chemosphere 2001; 43(4-7): 937-42. [http://dx.doi.org/10.1016/S0045-6535(00)00454-9] [PMID: 11372887]

[12] Eskenazi B, Mocarelli P, Warner M, *et al.* Serum dioxin concentrations and endometriosis: a cohort study in Seveso, Italy. Environ Health Perspect 2002; 110(7): 629-34. [http://dx.doi.org/10.1289/ehp.02110629] [PMID: 12117638]

[13] Eskenazi B, Mocarelli P, Warner M, *et al.* Relationship of serum TCDD concentrations and age at exposure of female residents of Seveso, Italy. Environ Health Perspect 2004; 112(1): 22-7. [http://dx.doi.org/10.1289/ehp.6573] [PMID: 14698926]

[14] Warner M, Eskenazi B, Patterson DG, *et al.* Dioxin-Like TEQ of women from the Seveso, Italy area by ID-HRGC/HRMS and CALUX. J Expo Anal Environ Epidemiol 2005; 15(4): 310-8.
[http://dx.doi.org/10.1038/sj.jea.7500407] [PMID: 15383834]

[15] Warner M, Eskenazi B, Olive DL, *et al.* Serum dioxin concentrations and quality of ovarian function in women of Seveso. Environ Health Perspect 2007; 115(3): 336-40.
[http://dx.doi.org/10.1289/ehp.9667] [PMID: 17431480]

[16] Eskenazi B, Warner M, Samuels S, *et al.* Serum dioxin concentrations and risk of uterine leiomyoma in the Seveso. Women's Health Study Am J Epidemiol 2007; 166(1):79-87.

[17] Matilal S, Höpfl H. Accounting for the Bhopal disaster: footnotes and photographs. Account Audit Account J 2009; 22(6): 953-72.
[http://dx.doi.org/10.1108/09513570910980472]

[18] Eckerman I. Chemical Industry and Public Health Bhopal as an Example. Essay for the Master of Public Health Nordic School of Public Health. Göteborg, Sweden. Course N. MPH200124. Nordiska hälsovårdshögskolan. 22.01.2001. Accessed 28/07/2020. https://www.lakareformiljon.org/images/stories/dokument/2009/bhopal_gas_disaster.pdf

[19] The World's Worst Pollution Problems: Assessing Health Risks at Hazardous Waste Sites 2012. www.worstpolluted.org

[20] Gensburg LJ, Pantea C, Kielb C, Fitzgerald E, Stark A, Kim N. Cancer incidence among former Love Canal residents. Environ Health Perspect 2009; 117(8): 1265-71.
[http://dx.doi.org/10.1289/ehp.0800153] [PMID: 19672407]

[21] Kielb CL, Pantea CI, Gensburg LJ, *et al.* Concentrations of selected organochlorines and chlorobenzenes in the serum of former Love Canal residents, Niagara Falls, New York. Environ Res 2010; 110(3): 220-5.
[http://dx.doi.org/10.1016/j.envres.2009.11.004] [PMID: 20117765]

[22] Vianna NJ, Polan AK. Incidence of low birth weight among Love Canal residents. Science 1984; 226(4679): 1217-9.
[http://dx.doi.org/10.1126/science.6505690] [PMID: 6505690]

[23] Goldman LR, Paigen B, Magnant MM, Highland JH. Low birth weight, prematurity and birth defects in children living near the hazardous waste site Love Canal. Hazard Waste Hazard Mater 1985; (2)209-23.
[http://dx.doi.org/10.1089/hwm.1985.2.209]

[24] Thompson J, Anthony H. The health effects of waste incinerators. Brit Soc Ecol Med 2005; 15: 115-156.

[25] Air pollution concerns at smaller airports Filed in Air Pollution, Airports, Issues, News from the AEF Feb 21 2019. https://www.aef.org.uk/2019/02/21/air-pollution-concerns-at-smaller-airports/

[26] Irvine D, Budd L, Ison S, Kitching G. The environmental effects of peak hour air traffic congestion: The case of London Heathrow Airport Research in Transportation Economics, 2016; 55:67-7, Loughborough University, posted on 13.07.2016, 13:15 by Daniel Irvine Lucy Budd Stephen Ison Gareth Kitching©. 2016 Elsevier Ltd. Accessed https://repository.lboro.ac.uk/articles/The_environmental_effects_of_peak_hour_air_traffic_congestion_the_case_of_London_Heathrow_Airport/9451085 28/07/2020.

[27] Nichols TP, Leinster P, McIntyre AE, Lester JN, Perry R. A survey of air pollution in the vicinity of Heathrow airport (London). Sci Total Environ 1981; 19: 285-92.
[http://dx.doi.org/10.1016/0048-9697(81)90023-1]

[28] Masiol M, Harrison RM. Quantification of air quality impacts of London Heathrow Airport (UK) from 2005 to 2012. Atmos Environ 2015; 116: 308-19.
[http://dx.doi.org/10.1016/j.atmosenv.2015.06.048]

[29] Flanagan P. 'Huge nuclear dump near Sellafield 'will leak into the Irish sea, Experts describe the site, which is just kilometres from Ireland's east coast, as a 'slow motion Chernobyl''. Irish Mirror on line. 22/04/2014 (St https://www.irishmirror.ie/news/irish-news/huge-nuclear-dump-near-sellafield-3443341. Accessed 28/07/2020.

[30] Sellafield Environmental Protection Agency. https://www.epa.ie/radiation/monassess/sellafield/

[31] Oppenheimer A. Britain: Sellafield, Salmon, and The Irish Sea. Bulletin of the Atomic Scientists 01/09/2003; 59 issue: 5, page(s): 11-13 Sage Journals. https://journals.sagepub.com/doi/full/10.2968/059005004. Accessed 28/07/2020.

[32] Doman B. want the truth!' Fear and suspicion in Rouen after chemical plant blaze Issued on: 01/10/2019 - 19:12Modified: 02/10/2019 - 09:06. https://www.france24.com/en/20191001-france-rouen-chemical-plant-fire-lubrizol-pollution-philippe

[33] French factory fire could pollute Seine river, officials warn. 26 September 2019. https://www.bbc.com/news/world-europe-49839570

[34] Riana Bornman MS, Bouwman H. Environmental pollutants and diseases of sexual development in humans and wildlife in South Africa: harbingers of impact on overall health? Reprod Domest Anim 2012; 47 (Suppl. 4): 327-32.
[http://dx.doi.org/10.1111/j.1439-0531.2012.02094.x] [PMID: 22827388]

[35] Beard J. DDT and human health. Sci Total Environ 2006; 355(1-3): 78-89.
[http://dx.doi.org/10.1016/j.scitotenv.2005.02.022] [PMID: 15894351] [Review].

[36] Lovett RA. Oceans release DDT from decades ago 7 January 2010; Nature.
[http://dx.doi.org/10.1038/news.2010.4]

[37] World Health Organization and UNEP. HELI Health, Environment Linkages Website. Toxic Hazards https://www.who.int/heli/risks/toxics/chemicals/en/. Accessed 28/07/2020.

[38] Rahman MM. Insecticide substitutes for DDT to control mosquitoes may be causes of several diseases. Environ Sci Pollut Res Int 2013; 20(4): 2064-9.
[http://dx.doi.org/10.1007/s11356-012-1145-0] [PMID: 22956113]

[39] Lim X. Tainted water: the scientists tracing thousands of fluorinated chemicals in our environment. Nature 2019; 566(7742): 26-29.
[http://dx.doi.org/10.1038/d41586-019-00441-1]

[40] Luz AL, Anderson JK, Goodrum P, Durda J. Perfluorohexanoic acid toxicity, part I: Development of a chronic human health toxicity value for use in risk assessment. Regul Toxicol Pharmacol 2019; 103: 41-55.
[http://dx.doi.org/10.1016/j.yrtph.2019.01.019] [PMID: 30639337]

[41] Anderson JK, Luz AL, Goodrum P, Durda J. Perfluorohexanoic acid toxicity, part II: Application of human health toxicity value for risk characterization. Regul Toxicol Pharmacol 2019; 103: 10-20.
[http://dx.doi.org/10.1016/j.yrtph.2019.01.020] [PMID: 30634020]

[42] Xiao F. Emerging poly- and perfluoroalkyl substances in the aquatic environment: A review of current literature. Water Res 2017; 124(124): 482-95.
[http://dx.doi.org/10.1016/j.watres.2017.07.024] [PMID: 28800519]

[43] Corton JC, Peters JM, Klaunig JE. The PPARα-dependent rodent liver tumor response is not relevant to humans: addressing misconceptions. Arch Toxicol 2018; 92(1): 83-119.
[http://dx.doi.org/10.1007/s00204-017-2094-7] [PMID: 29197930]

Miriam Kennet
Chartered Institute of Purchasing and Supply- MCIPS
Alumna of the Month South Bank University
London
Editor Green Academic Journal
Director CEO The Green Economics Institute
Head of United Nations Delegation to the COP Climate Conferences

List of Contributors

Amal I. Hassan	Radioisotope Department, Nuclear Research Center, Atomic Energy Authority, Dokki 12311, Giza, Egypt
Arturo Hermann	Italian National Institute of Statistics (ISTAT), Viale Liegi 13 Rome, Italy
Carla Di Stefano	Department of Hematology, "Tor Vergata", University, Viale Oxford 81, 00133 Rome, Italy
Chuaqui Jorge	University of Valparaiso, Chile
Debra R. Wilson	Austin Peay State University, USA
Gabriella Marfe	Department of Scienze e Tecnologie Ambientali, Biologiche e Farmaceutiche, University of Campania "Luigi Vanvitelli," *via* Vivaldi 43, Caserta 81100, Italy
Hosam M. Saleh	Radioisotope Department, Nuclear Research Center, Atomic Energy Authority, Dokki 12311, Giza, Egypt
James G. Linn	Optimal Solutions in Healthcare and International Development, USA
Luigi Montano	Chief of Andrology Unit and Lifestyle Medicine - Local Health Authority (ASL) Salerno, EcoFoodFertility Project Coordination Unit, Oliveto Citra (SA), Italy
Rajiv Ganguly	Department of Civil Engineering, Jaypee, University of Information Technology, Waknaghat, District Solan, Himachal Pradesh 173234, India
Rishi Rana	Department of Civil Engineering, Jaypee, University of Information Technology, Waknaghat, District Solan, Himachal Pradesh 173234, India
Stefania Perna	Department of Scienze e Tecnologie Ambientali, Biologiche e Farmaceutiche, University of Campania "Luigi Vanvitelli," *via* Vivaldi 43, Caserta 81100, Italy
Sunil Jayant Kulkarni	Gharda Institute of Technology, Lavel, Maharashtra, 415708, India
Thabo T. Fako	Vice Chancellor University of Botswana Gaborone, Botswana

CHAPTER 1

Introduction

Gabriella Marfe[1,*], **Stefania Perna**[1] and **Carla Di Stefano**[2]

[1] *Department of Scienze e Tecnologie Ambientali, Biologiche e Farmaceutiche, University of Campania "Luigi Vanvitelli," via Vivaldi 43, Caserta 81100, Italy*

[2] *Department of Hematology, "Tor Vergata" University, Viale Oxford 81, 00133 Rome, Italy*

Abstract: Waste pollution, with its harmful health risks, is one of the critical issues present in each country. Waste generation rates are growing around the world every year, creating a need for more sustainable waste management strategies. Random disposal of wastes is creating a very complicated situation that involves all world populations. Although several studies have been conducted to investigate links between waste pollution and cancer; such evidence about this correlation becomes very difficult to demonstrate. The information about the effects of environmental exposure in the population is insufficient, and it is very challenging to establish the impact of human health. In this scenario, interdisciplinary research can play an important role to better understand the relationship between the risks of human health and waste pollution.

Keywords: Adverse Health Impact, Diseases, Exposure Effect, Pollution, Waste.

INTRODUCTION

Hazardous waste management is becoming a significant challenge for sustainable development for all nations. In Agenda 21, the Rio Declaration on Environment and Development, it was stated that environmentally sound management (ESM) of wastes was among the most important issue to improve the quality of the earth's environment [1, 2]. The United Nation's Sustainable Development Goals (SDGs) have set out some goals and targets linked to waste management [3]. Environmental contamination due to hazardouswaste mismanagement is a global issue. But today, hazardous waste is still dumped in an unsafe manner poisoning our environment and causing severe risks to the health of both humans and animals. Pollution from plastic bags is associated with several environmental problems: the waste blocks gutters and drains, resulting in stormwater problems. Furthermore, livestock can die after plastic consumption. For example, the plastic bags are often non-biodegradable and for this reason, soil productivity can decr-

* **Corresponding author Gabriella Marfe:** Department of Scienze e Tecnologie Ambientali, Biologiche e Farmaceutiche, University of Campania "Luigi Vanvitelli," *via* Vivaldi 43, Caserta 81100, Italy; Tel: +39 0823 275104; Fax: +39 0823 274813; E-mail: gabmarfe@alice.it

ease because of their presence in agricultural welds. Moreover, the burning of the plastic is very dangerous since toxic gases such as furan and dioxin are released along with residues (including lead and cadmium) that are deposited on the ground [4, 5]. Different studies found high levels of heavy metals at the dumpsites, and additionally, the residents, living close to such sites, had respiratory ailments and high lead levels in the blood (the internationally acceptable level 10 µg/dl) [6].

Four steps in the risk assessment process are recommended by the US Environmental Protection Agency (USEPA) and The National Academy of Science. The first step aims to characterize the hazard at a specific site and, then it is necessary to compare each contaminant with Risk-Based Concentrations [7, 8]. Such hazard characterization is used to compile a list of pollutants that can be dangerous for environmental health [6]. The second step aims to identify the toxicity or dose-response assessment. In most cases, the EPA's Integrated Risk Information System (IRIS) is utilized to identify a Reference Dose (RfD) for non-cancer effects, and/or a Cancer Slope Factor (CSF) for cancer effects. There are five categories used by the USEPA in order to classify the carcinogenic potential of a specific chemical.

These include:

1. Class A—evidence of carcinogenicity in humans,
2. Class B—probable human carcinogen (limited evidence in humans and adequate evidence in animals),
3. Class C—possible human carcinogen (sufficient evidence in animals),
4. Class D—not classifiable for human carcinogenicity, and
5. Class E— evidence for non-carcinogenicity in humans.

The cancer slope factor, used by the USEPA, represents the carcinogenic potency of different chemicals. For example, methylene chloride (Class B2) is considered a weak carcinogen (Class B2) and has an oral cancer slope factor of 7.5×10^{-3} (mg/kg)/day, while vinyl chloride is considered a strong human carcinogen (Class A) and has an oral cancer slope factor of 7.2×10^{-1} [7, 8].

The third step in this process aims to evaluate the exposure. It is essential to identify all exposure pathways at a site for both on-site workers and off-site residents. This process considers exposure to soil, air, groundwater, surface water, sediment, or food products that may be contaminated by chemicals derived from the site. For each exposure pathway, the concentration of each dangerous contaminant is calculated in specific media. In this way, it is possible to obtain a mean value, an upper 90th percentile value, or a maximum value depending on the quantity and quality of existing data. Furthermore, such values are utilized to calculate a Cumulative Daily Intake for each exposure pathway and each dangerous contaminant. In this regard, among different sources in the exposure assessment, the intake variables for contaminated media, the estimate of chemical concentrations in the media, and the absorption rate from several exposure pathways must be considered [9].

The last stage in the risk assessment provides the characterization of both the non-cancer and cancer risks. The non-cancer risk depends on a sum of the Hazard Quotient, or a value obtained by dividing the Cumulative Daily Intake by the Reference Dose. Furthermore, contaminant concentrations are estimated acceptable when the daily value is not higher than the Reference Dose (or No Observable Adverse Effect Level). Hazard Quotients for each chemical and each exposure pathway are summed. The Lifetime Cancer Risk is estimated as a product of the Cancer Slope Factor and the cumulative Daily Intake. Residential sites are considered safe when the sum of Lifetime Cancer Risk for all chemicals and all exposure pathways is not superior in one in-one million[1]. The classification of sites and evaluating acceptable levels for clean-up is estimated by risk calculations. However, different sources of uncertainty exist in each step of this process [10 - 13].

Although humans are constantly exposed to different toxic wastes, it is difficult to study the cancer rate, considering the variability of these exposures and the variability of the human population. In addition, the tumor can develop after 20 years from exposure to toxic substances. Animal models are a good tool to estimate the genotoxic interaction with chemical mixtures because of their uniform genetics [13, 14]. The association between animal data and epidemiological data gives an accurate prediction about the considered compound(s) [14 - 16]. Epidemiological data can contribute significantly to studies on human carcinogenicity, especially occupational exposure studies 1 [4 - 17]. Within such studies, it is important to consider interspecies differences such as age, sex, rate of metabolic processes, *etc.* Besides, exposure to human populations is often highly variable. Biomarkers, such as metabolites and DNA adduct formation, may be monitored in exposed people to determine exposure and response relationships as accurately as possible [14, 18 - 20]. For this reason, there is a poor understanding of the long-term effects of chemicals in such waste sites, although several studies highlight negative health impact due to waste management practices, in particular hazardous waste among developed, less developed and developing countries [21 - 24]. Today, few landfills in poorer countries could possess environmental standards accepted by industrialized nations. Furthermore, it would be difficult to redevelop them since the financial resources of these countries are limited for their remediation. Another important environmental issue is gas release by decomposing garbage [25, 26] since such gases contribute to increased greenhouse gas effect and climate change. In this context, we must remember the most famous examples of groundwater pollution in the USA, the Love Canal tragedy in Niagara Falls, New York. During World War II (1940), Love Canal was used to dump waste and then, in 1942, Hooker Chemical purchased the land and disposed of about 21,000 tons of hazardous chemical waste (exachlorocyclohexanes, benzylchlorides, organic sulfur compounds, chlorobenzenes, and sodium sulfide/sulfhydrates). After eleven years (1953), Hooker sold the land to the Niagara Falls School Board. In the late '50s, the School Board built a public school and sold for housing. From the 1950s to the late 1970s, residents experienced several health problems such as high rates of miscarriages, birth defects, and chromosome damage. In August 1978, the lead paragraph of a front-page story in the New York Times read:

Niagara Falls N.Y. twenty five years after the Hooker Chemical Company stopped using the Love Canal here as an industrial dump, 82 different compounds, 11 of them suspected carcinogens, have been percolating upward through the soil, their drum containers rotting and leaching their contents into the backyards and basements of 100 homes and public school build on the banks of the canal.

At the end of 1978, President Carter declared a state of emergency at Love Canal and Congress created the Comprehensive Environmental Response, Compensation and Liability Act (CERCLA), informally called Superfund in 1980. This act allowed the Environmental Protection Agency (EPA) to clean up numerous contaminated sites. Nevertheless, over 1,000 hazardous waste sites are still in the process of cleaning with an increased risk to both human health and the environment [27 - 32].

In 2007, The US Environmental Protection Agency's National Priority List (NPL) recognized 1240 hazardous waste sites, among them, 157 are federal facilities. Furthermore, EPA estimated that 41 million people were living close to NPL sites (about a 4-mile radius). In Europe, 342,000 contaminated sites were identified in 2014 (5.7 per 10,000 inhabitants) [33].

The increase in hazardous waste generation in China is correlated with its rapid economic growth. In one study, the authors found that Chinese workers from an e-waste dismantling site had high serum levels of thyroid-stimulating hormone and polybrominated diphenyl ethers. Furthermore, these workers presented an elevated frequency of micronuclei (MN) in peripheral lymphocytes that are generally used as biomarkers of genetic damage in humans with known exposure to genotoxic agents [34]. Furthermore, different studies reported many cases of death and illness around areas exposed to hazardous waste materials, including the so-called 'cancer villages' of China. In such villages, the residents have a high cancer rate than the expected rate. Such results could be directly linked to their exposure to cadmium and mercury released by the informal recycling of e-waste [35 - 40]. During the 1990s, some reports pointed out the relationship between an increased cancer rate and pollution in China. In the early 2000s, international media and different NGOs (Non-Governmental Organizations) started to document the existence of cancer villages [41 - 43]. These villages are associated with the chemical industry, paper factories, or resource extraction and processing and most of them are located in Eastern China [41 - 46]. One famous example is the Huai River Delta, where many small factories dump their waste directly into the river, poisoning water, and in turn, causing large-scale health issues [47, 48]. The death rates in these villages exceed the national average: it means that China is paying breakneck economic growth with a tremendous cost of human lives [49].

Many Asian countries are characterized by improper waste management. The urbanization, industrialization, and the rapid economic growth have caused an uncontrolled increase of municipal solid waste, also changing its contents. Today, Singapore, Korea and Japan have reached high standards in their municipal waste management. For instance, the Republic of Korea recycled or composted almost 60% of its municipal waste, surpassing Europe [50].

The unsafe disposal and its potential impact on surface and groundwater are the principal concerns correlated with solid waste in other Asian countries. For instance, in Vietnam, although most water results polluted, it is used for drinking, domestic purposes, agriculture, and aquaculture. In this regard, the pollution of these essential resources is likely to have severe health risks for the local population. Therefore, the mortality and morbidity rates have high levels of diseases in correlation with the water supply and sanitation. The drinking-water resources can be contaminated when untreated waste, along with the incidental discharge of oil and other chemicals, is poured directly into rivers or underground water supply systems. In this context, the local populations start to suffer from heart disease, infections of the respiratory and digestive system, and dermatitis [51]. In Taiwan, spatial autocorrelation analysis identified hot spots for various cancers in females in areas with high levels of environmental exposures to arsenic, nickel, and chromium [52]. A recently published paper has reported an increased incidence of different cancers in Lebanon between 2003 and 2007 [53]. Furthermore, the same paper has calculated based on previous estimates that the rate of cancer for both males and females will increase by 2020: from 0.16% in 2008 to 0.23% in 2020 for men and from 0.19% to 0.28% for women. In this country, the municipal solid waste disposal continued to rely on landfills (about 53% of generated municipal waste). Such wastes are collected and disposed of in uncontrolled dumpsites. Generally, methane gas, released from the internal parts of these dumpsites, causes fires, which, in turn, lead to significant fall of polychlorinated Dibenzo- P-Dioxins and polychlorinated Dibenzofurans (PCDDs/PCDFs) emissions [54, 55]. Several of the dumpsites are located close to rivers and streams, as well as urban communities. Furthermore, both mercury (10%) and polycyclic aromatic hydrocarbons (PAH) (40%) emissions derive from open burning. Such pollution is correlated with different diseases such as lung and neurological diseases, heart attacks and some cancers [56]. Other authors underlined a dramatic situation in Malaysia where healthcare waste was generally treated by incineration. This technology causes high levels of substances such as dioxin and furan that are involved in the development of several diseases (liver failure and cancer) [57]. Many toxic waste sites are also identified in some countries of Africa such as Ethiopia, Nigeria, Tanzania, South Africa, Ghana and Somalia. Both the public and private sectors of waste management in many African countries do not work well and the illegal dumping of different kinds of waste is a common practice. In addition, funds for waste management are minimal that, in turn, causes an inefficiency in the public service [58, 59]. For instance, in Ethiopia, healthcare waste was generally treated by incineration in both the public and private hospitals. Generally, the public hospitals had open pits in their backyards that were used for the final disposal as burial or open incineration, while the private hospitals were mixing their healthcare waste with the municipal solid waste. [60]. Similar situations have been reported in Nigeria [61 - 63], Botswana [64], and Cameroon [65]. In Tanzania, several contaminated sites contain hazardous waste and they are not cleaned up because of a lack of financial resources. According to a 1998 study by the National Environment Management Council, there are more than 300 contaminated sites in Tanzania [66]. Two papers reported that different factors affected the improper e-waste management in Ghana, such as a lack of e-waste-specific legislation, an inadequate infrastructure, and low public awareness and education [67]. In particular, the second study showed that the concentrations of some heavy metals at the Agbogbloshie (e-waste processing site in Accra, Ghana) were high when compared with the regulatory limits of both Dutch and Canadian Soil Quality and Guidance Values. Furthermore, such concentrations were also elevated in the main burning and dismantling sites and in the school, residential, recreational, clinic, farm and work-ship area [68]. Finally, a very well written review underlined that urban effluent discharges (including domestic and industrial effluent and/or urban stormwater run-off) can decrease fertility and increase cancer incidence in South Africa [69].

Very little is known about hazardous waste management in Russia and cities such as Leningrad Oblast, the City of St Petersburg and Kaliningrad Oblast. The only available information is the annual waste reports to Rostechnadzor from companies subject to federal control ("http://helcom.fi/helcom-at-work/projects/completed%20projects/balthazar/hazardous-waste"/hazardous-waste).

The lack of environmentally sound hazardous waste treatment technology, a poor collection of hazardous waste and inefficient enforcement of legislation are the main issues in Russia. In this regard, information about the amounts and types of hazardous waste are incomplete for environmental authorities. Rostechnadzor reports provide data on matters concerning hazardous waste generation and storage and PCB and mercury-containing waste [70, 71]. The design of sustainable and environmentally management facilities needs to have information on types and quantities of hazardous waste. In Russia, it is possible to collect hazardous waste in association with other waste from private citizens or households. Moreover, in one study, some authors found a high cancer incidence rate in many areas of Russia [72]. Furthermore, some studies in other countries reported some reproductive and cancer outcomes in populations living in proximity of landfills and incinerators [73 - 75].

Another story is "A toxic legacy. Illegal dumping of toxic waste in the Italian Campania" has been blamed for high rates of ill health in the region [76]. In the mid1980s, this Italian region has been used extensively to bury tons of toxic waste derived from Italian and foreign companies. According to many reports released by specialized international agencies such as the United Nations Environmental Program (UNEP), Greenpeace and other international environmentalist organizations, the impact of the toxic wastes dumping has compromised human health and quality of environment. In 2004, the scientific journal Lancet Oncology published a study of Kathryn Senior and Alfredo Mazza, entitled "Triangle of death linked to waste crisis". The authors found a significant rise of cancer rates in an area including the towns of Nola, Marigliano and Acerra (east of Naples), between 1994 and 2000 (Figs. **1** and **2**) [77]. Although numerous epidemiological studies have shown the high incidence of cancer in Campania, new research designs and procedures have to incorporate biological/biochemical tools.

Now an established heart consultant who has published further studies into the consequences of hazardous waste, Mazza admits it is impossible to prove precise links between toxic materials, tumours and congenital malformations. But he believes they are only just beginning to see the full scale of health problems.

'We are living in the Triangle of Death. We do not know how many areas are affected, how bad the damage will be or how long it will last.'

(https://www.telegraph.co.uk/news/0/mafia-toxic-waste-and-a-deadly-c-ver-up-in-an-italian-paradise-t/).

Moreover, in a recent article, Giordano *et al.* found high blood concentrations of heavy metals in 95 patients with different cancers, residing in in the eastern area of the Campania region (different municipalities, including Giugliano in Campania, where many illegal waste disposal sites have previously been documented) in comparison with healthy individuals [78].

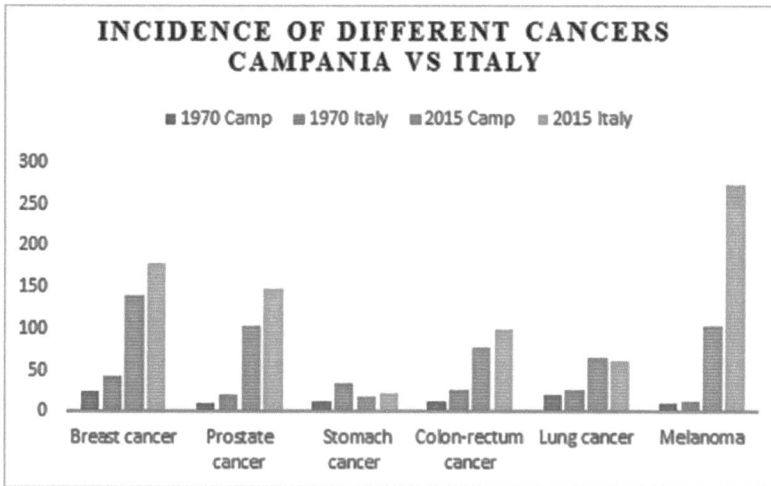

(http://www.tumori.net/banche_dati/tumori/query.php)

Fig. (1). Cancer incidence of different cancer Campania *vs.* Italy (http://www.tumori.net/banche_dati/ tumori/query.php).

Today, the literature highlighted a relationship between occupational or accidental exposure and health impact. Many studies, conducted on both animal and human models, showed that substances, such as cadmium, arsenic, chromium, nickel, dioxins and Polycyclic Aromatic Hydrocarbons (PAHs) can be considered carcinogenic. In addition, many of these substances can cause other toxic effects (depending on exposure level and duration) on the central nervous system, liver, kidneys, heart, lungs, skin, reproduction, *etc.* [79 - 81]. Another problem has been documented by recent monitoring of municipal waste-water effluent, urban surface water and biota that reported the presence of groups of unrecognized contaminants called contaminants of emerging concern (CECs) [82 - 84]. For other pollutants, including SO_2 and PM10, multiple studies have shown the adverse effects on morbidity and mortality at background levels of exposure in susceptible groups such as the elderly. Chemical substances such as dioxins and organochlorines may accumulate in fat-rich tissues and be correlated with reproductive or endocrine-disrupting endpoints. Recent articles indicated that more than half of cancer cases and 60% of deaths occur in developing countries [85]. In 2007, a WHO report on waste and health concluded:

"Despite the methodological limitations, the scientific literature on the health effects of landfills provides some indication of the association between residing near a landfill site and adverse health effects. The evidence, somewhat stronger for reproductive outcomes than for cancer, is not sufficient to establish the causality of the association. However, in consideration of the large proportion of population potentially exposed to landfills in many European countries and of the low power of the studies to find a real risk, the potential health implications cannot be dismissed" [86].

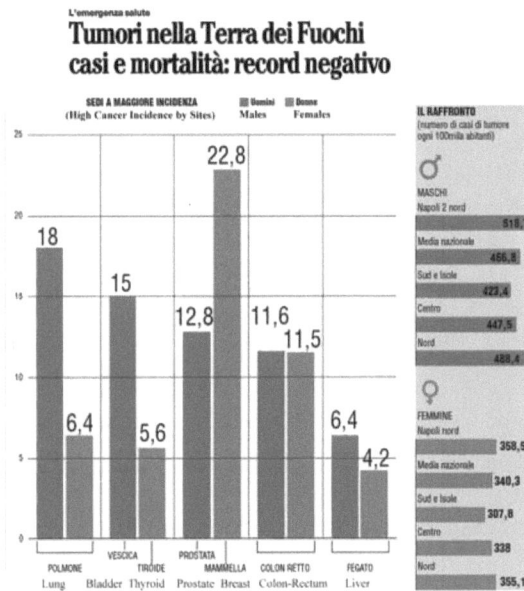

Fig. (2). Cancer incidence of liver, breast, lung, bladder, colon-rectum, prostate and thyroid in both sexes when compared with regional and national average, provided from ASL 2 Nord Naples (Mautone Ettore. Il Mattino 1-June 2017 www.ilmattino.it).

On a global scale, billions of tons of waste are generated every year and the health issues correlated with the waste disposal are augmenting in different countries. Today, it is important to invest in modern technologies for waste management and training and education to reduce the impact on human health. Such issue needs to be tackled on different ways, *i.e.* (i) Incentives for waste reduction, waste prevention, recycling, and composting, (ii) increased costs of waste management to consumer products, (iii) raising public awareness in waste handling at local and regional level, (iv) public health surveillance, and (v) the use of biomarker epidemiology techniques in future research. Across the world, the lack of accountability for the harm to the environment and public health caused by inadequate toxic waste management undermines the environmental and health rights of citizens. The different issues in this situation are: (1) a lack of community awareness and responsibility towards waste; (2) the lack of political commitment to address the toxic waste problem; (3) the scarcity of resources including finance, equipment, personnel and data for waste planning; (4) lack of enforcement of existing regulations on toxic solid waste and urban environmental management in general. Until these fundamental requirements are met, the adverse effects of an improper waste will affect the environment and human health.

Today we're dumping 70 million tons of global-warming pollution into the environment, and tomorrow we will dump more, and there is no effective worldwide response. Until we start sharply reducing global-warming pollution, I will feel that I have failed.
Al Gore

"We must remember the Newton's second law of motion: "For every action, there is an equal and opposite reaction. Today, nature is reacting in a strong manner to highly significant pollution of soil, water and air. All such pollution is done only for the interests of a few people in the name of huge economic profits. Climate change is already occurring, and the denial will not stop it. Each of us must do everything in our power to allow the survival of our future generations."
Gabriella Marfe

NOTES

[1]Cancer Potency Factor" or "Unit Risk" indicates that a) the carcinogenic potency has not been studied. Studies of their carcinogenic potency did not show a dose-related increased in cancer incidence, or some evidence of carcinogenic potency has been observed but the quality of the studies or the data does not allow quantitative estimation of carcinogenic potency. The dose associated with an increased cancer risk of one-in-one million may be calculated from the cancer potency factor (1×10^{-6} dose = 1×10^{-6}/cancer potency factor). The air concentration associated with an increased cancer risk of one-in-one million may be calculated from the unit risk (1×10^{-6} air concentration = 1×10^{-6}/unit risk).

CONSENT FOR PUBLICATION

Not applicable.

CONFLICT OF INTEREST

The authors confirm that the contents of this chapter have no conflict of interest.

ACKNOWLEDGEMENTS

Declare none.

REFERENCES

[1]   United Nations Conference on Environment & Development. Rio de Janerio, Brazil. 1992.3 to 14 June; *AGENDA 21*.

[2]   UNDESA. Agenda 21- Chapter 21 Environmentally Sound Management of Solid Wastes and Sewage-related Issues. Division for Sustainable Development, United Nations Department of Economic and Social Affairs 2005. (Available online at: http://www.un.org/esa/sustdev/documents/agenda21/ index.htm).

[3]   Transforming our world: the 2030 Agenda for Sustainable Development UN 2015. https://sustainable development.un.org/post2015/transformingourworld

[4]   Njeru J. The urban political ecology of plastic bag waste problem in Nairobi, Kenya. Geoforum 2006; 37(6)
        [http://dx.doi.org/10.1016/j.geoforum.2006.03.003]

[5]   UNEP. 2007.Annual Report

[6]   United States Environmental Protection Agency (USEPA). 2006a.EPA region 3 risk based concentration table: technical background information http://www.epa.gov/reg3hwmd/risk/ human/info/tech.htm

[7]   United States Environmental Protection Agency (USEPA). Integrated risk information system (IRIS) toxicological review: dichloromethane http://www.epa.gov/iris, 1995.

[8]   United States Environmental Protection Agency (USEPA). Integrate;d risk information system (IRIS) toxicological review: vinyl chloride http://www.epa.gov/iris, 2000a.

[9]   US EPA. Peer Review Handbook EPA/100/B-06/002. Washington, DC: US Environmental Protection Agency 2006. b

[10]   United States Environmental Protection Agency (USEPA). 1994b.Integrated risk information system (IRIS) toxicological review: benzo(a)pyrene (BaP) http://www.epa.gov/iris

[11]   United States Environmental Protection Agency (USEPA). 1986.Guidelines for the health risk assessment of chemical mixtures http://www.epa.gov/ncea/raf/pdfs/chem_mix/chemmix?1986.pdf

[12]   United States Environmental Protection Agency (USEPA). 2000b.Second five-year review report for Libby   ground   water   site,   Libby,   Lincoln   County,   Montana http://www.epa.gov/superfund/action/law/sara.htm

[13]   Barrett JC. Mechanisms for species differences in receptor-mediated carcinogenesis. Mutat Res 1995; 333(1-2): 189-202.
        [http://dx.doi.org/10.1016/0027-5107(95)00145-X] [PMID: 8538627]

[14]   United States Department of Health and Human Services (DHHS). 11th Report on carcinogens. Prepared for the National Institute of Environmental Health Sciences.. Durham, NC: Constella Group, Incorporated 2004.

[15]   Wogan GN, Hecht SS, Felton JS, Conney AH, Loeb LA. Environmental and chemical carcinogenesis. Semin Cancer Biol 2004; 14(6): 473-86.
        [http://dx.doi.org/10.1016/j.semcancer.2004.06.010] [PMID: 15489140]

[16]   Miller JA, Miller EC. Chemical carcinogenesis: mechanisms and approaches to its control. J Natl Cancer Inst 1971; 47(3): V-XIV.
        [PMID: 5157587]

[17]   Krewski D, Thomas RD. Carcinogenic mixtures. Risk Anal 1992; 12(1): 105-13.

[http://dx.doi.org/10.1111/j.1539-6924.1992.tb01313.x] [PMID: 1574610]

[18] Grimmer G, Jacob J, Dettbarn G, Naujack KW. Determination of urinary metabolites of polycyclic aromatic hydrocarbons (PAH) for the risk assessment of PAH-exposed workers. Int Arch Occup Environ Health 1997; 69(4): 231-9.
[http://dx.doi.org/10.1007/s004200050141] [PMID: 9137996]

[19] Malkin R, Kiefer M, Tolos W. 1-Hydroxypyrene levels in coal-handling workers at a coke oven. J Occup Environ Med 1996; 38(11): 1141-4.
[http://dx.doi.org/10.1097/00043764-199611000-00014] [PMID: 8941904]

[20] Melber C, Kielhorn J, Mangelsdorf I. Concise International Chemical Assessment Document 62: Coal Tar Creosote. Geneva, Switzerland: World Health Organization 2004.

[21] Vrijheid M. Health effects of residence near hazardous waste landfill sites: a review of epidemiologic literature. Environ Health Perspect 2000; 108 (Suppl. 1): 101-12.
[PMID: 10698726]

[22] Rushton L. Health hazards and waste management. Br Med Bull 2003; 68: 183-97.
[http://dx.doi.org/10.1093/bmb/ldg034] [PMID: 14757717]

[23] Franchini M, Rial M, Buiatti E, Bianchi F. Health effects of exposure to waste incinerator emissions:a review of epidemiological studies. Ann Ist Super Sanita 2004; 40(1): 101-15.
[PMID: 15269458]

[24] Saunders P. A systematic review of the evidence of an increased risk of adverse birth outcomes in populations living in the vicinity of landfill waste disposal sites.Population health and waste management: scientific data and policy options. Report of a WHO workshop Rome, Italy, 29-30 March 2007 In: Mitis F, Martuzzi M, Eds. WHO, Regional Office for Europe, Copenhagen. 2007; 16: pp. 25-7.

[25] Cointreau-Levin S Occupational and Environmental Health Issues of Solid Waste Management: Special Emphasis on Middle and Lower – Income Countries. (Draft) World Bank 1997.http://www.ilsr.org/recycling/other/dctransfer/ochealth.pd

[26] Pervez A, Ahmade K. Impact of Solid Waste on Health and The Environment International. Int J Sustain Devel Green Econ (IJSDGE) 2013; 2: 165-8.

[27] The World's Worst Pollution Problems: Assessing Health Risks at Hazardous Waste Sites 2012.www.worstpolluted.org

[28] http://depts.washington.edu/envir202/Readings/Reading05.pdf

[29] Gensburg LJ, Pantea C, Kielb C, Fitzgerald E, Stark A, Kim N. Cancer incidence among former Love Canal residents. Environ Health Perspect 2009; 117(8): 1265-71.
[http://dx.doi.org/10.1289/ehp.0800153] [PMID: 19672407]

[30] Kielb CL, Pantea CI, Gensburg LJ, *et al.* Concentrations of selected organochlorines and chlorobenzenes in the serum of former Love Canal residents, Niagara Falls, New York. Environ Res 2010; 110(3): 220-5.
[http://dx.doi.org/10.1016/j.envres.2009.11.004] [PMID: 20117765]

[31] Vianna NJ, Polan AK. Incidence of low birth weight among Love Canal residents. Science 1984; 226(4679): 1217-9.
[http://dx.doi.org/10.1126/science.6505690] [PMID: 6505690]

[32] Goldman LR, Paigen B, Magnant MM, Highland JH. Low birth weight, prematurity and birth defects in children living near the hazardous waste site Love Canal. Hazard Waste Hazard Mater 1985; (2): 209-2239.
[http://dx.doi.org/10.1089/hwm.1985.2.209]

[33] Fazzo L, Minichilli F, Santoro M, *et al.* Hazardous waste and health impact: a systematic review of the scientific literature. Environ Health 2017; 11;16(1): 107.

[http://dx.doi.org/10.1186/s12940-017-0311-8]

[34]   Yuan J, Chen L, Chen D, *et al.* Elevated serum polybrominated diphenyl ethers and thyroid-stimulating hormone associated with lymphocytic micronuclei in Chinese workers from an e-waste dismantling site. Environ Sci Technol 2008; 42: 2195-200.
        [http://dx.doi.org/10.1021/es702295f]

[35]   Luo P, Bao LJ, Li SM, Zeng EY. Size-dependent distribution and inhalation cancer risk of particle-bound polycyclic aromatic hydrocarbons at a typical e-waste recycling and an urban site. Environ Pollut 2015; 200: 10-5.
        [http://dx.doi.org/10.1016/j.envpol.2015.02.007] [PMID: 25686883]

[36]   Lau WK, Liang P, Man YB, Chung SS, Wong MH. Human health risk assessment based on trace metals in suspended air particulates, surface dust, and floor dust from e-waste recycling workshops in Hong Kong, China. Environ Sci Pollut Res Int 2014; 21(5): 3813-25.
        [http://dx.doi.org/10.1007/s11356-013-2372-8] [PMID: 24288065]

[37]   Shi J, Zheng GJ, Wong MH, *et al.* Health risks of polycyclic aromatic hydrocarbons *via* fish consumption in Haimen bay (China), downstream of an e-waste recycling site (Guiyu). Environ Res 2016; 147: 233-40.
        [http://dx.doi.org/10.1016/j.envres.2016.01.036] [PMID: 26897061]

[38]   Song Q, Li J. A review on human health consequences of metals exposure to e-waste in China. Environ Pollut 2015; 196(196): 450-61.
        [http://dx.doi.org/10.1016/j.envpol.2014.11.004] [PMID: 25468213]

[39]   Zheng J, Chen KH, Yan X, *et al.* Heavy metals in food, house dust, and water from an e-waste recycling area in South China and the potential risk to human health. Ecotoxicol Environ Saf 2013; 96: 205-12.
        [http://dx.doi.org/10.1016/j.ecoenv.2013.06.017] [PMID: 23849468]

[40]   He X, Jing Y, Wang J, *et al.* Significant accumulation of persistent organic pollutants and dysregulation in multiple DNA damage repair pathways in the electronic-waste-exposed populations. Environ Res 2015; 137: 458-66.
        [http://dx.doi.org/10.1016/j.envres.2014.11.018] [PMID: 25679774]

[41]   Liu L. Made in China: Cancer Villages. Environment Science and Policy for Sustainable Development http://www.environmentmagazine.org/Archives/Back%20Issues/March-April%202010/made-in-china -full.htmlApril2015.

[42]   McBeath JH. Environmental Pollution, Cancer Villages and the State's Response. American Association   for   Chinese   Studies 2014.http://aacs.ccny.cuny.edu/2014conference/Papers/Jenifer%20McBeath.pdf(viewed April 2015)

[43]   Watt J. China's 'cancer villages' reveal dark side of economic boom The Guardian Xinglong [online] 2010. Available at: http://www.theguardian.com/environment/2010/ jun/07/china-cancer-villages industrial- pollution (viewed April 2015)

[44]   Lora-Wainwright A. Health, illness and welfare in rural China, School of Interdisciplinary Area Studies,   Social   Sciences   Division,   [online]   2015.   Available   at: http://www.area-studies.ox.ac.uk/health-illness-and-welfare-rural-china-anna-lora-wainwright(viewed April 2015)

[45]   Lora-Wainwright A. 2015, Health, illness and welfare in rural China, School of Interdisciplinary Area Studies,   Social   Sciences   Division,   [online]   Available   at: http://www.area-studies.ox.ac.uk/health-illness-and-welfare-rural-china-anna-lora-wainwright (viewed April, 2015)

[46]   Tseming Y. China's Cancer Villages Forthcoming, The Cambridge Handbook on Environmental Justice   and   Sustainable   Development   Sumudu   Atapattu,   Carmen   G,   Sara   Seck. 2018.https://ssrn.com/abstract=3277638

[47]   Elizabeth C. Economy, The River Runs Black: The Environmental Challenge to China's Future. Ithaca: Cornell University Press 2004.

[48]   Economy EC. The River Runs Black. USA: Cornell U. Press 2004.

[49]   Rappa A. The River Runs Black. USA: Cornell U. Press 2004.

[50]   Sustainable development in the European Union Eurostat 2015.

[51]   Waste Management: Issues and Challenges in Asia APO. 2007.

[52]   Chiang CT, Lian IeB, Chang YF, Chang TK. Geospatial disparities and the underlying causes of major cancers for women in Taiwan. Int J Environ Res Public Health 2014; 11(6): 5613-27.
[http://dx.doi.org/10.3390/ijerph110605613] [PMID: 24865397]

[53]   Azar S, Azar S. Waste related pollutions and their potential effect on cancer incidences in lebanon. J Environ Prot (Irvine Calif) 2016; 7: 778-83.
[http://dx.doi.org/10.4236/jep.2016.76070]

[54]   Shamseddine A, Saleh A, Charafeddine M, *et al.* Cancer trends in Lebanon: a review of incidence rates for the period of 2003–2008 and projections until 2018. Popul Health Metrics 2014; 12: 4.
[http://dx.doi.org/10.1186/1478-7954-12-4]

[55]   Lemieux PM, Lutes CC, Abbott JA, Aldous KM. Emissions of polychlorinated dibenzo-p-dioxins and polychlorinated dibenzofurans from the open burning of household waste in barrels. Environ Sci Technol 2000; 34: 377-84.
[http://dx.doi.org/10.1021/es990465t]

[56]   Hilal N, Fadlallah R, Jamal D, El-Jardali F. K2P Evidence Summary: Approaching the Waste Crisis in Lebanon: Consequences and Insights into Solutions. Knowledge to Policy (K2P) Center. Beirut, Lebanon 2015.

[57]   Ghasemi K, Yusuff RN. Advantages and disadvantages of healthcare waste treatment and disposal alternatives: Malaysian scenario. Pol J Environ Stud 2016; 25: 17-25.
[http://dx.doi.org/10.15244/pjoes/59322]

[58]   Zurbrugg C, Ahmed R. Enhancing Community Motivation and Participation in Solid Waste Management Department of Water and Sanitation in Developing Countries (SANDEC) at the Swiss Federal Institute for Environmental Science and Technology (EAWAG) 2000 January; SANDEC News, No. 4

[59]   Hussein BM. THE EVIDENCE OF TOXIC AND RADIOACTIVE WASTES DUMPING IN SOMALIA AND ITS IMPACT ON THE ENJOYMENT OF HUMAN RIGHTS: A CASE STUDY.Geneva, 8th of June, 2010 (https://fliphtml5.com/fzui/btjb/basic).

[60]   Tesfahun E. Tesfahun E. A study of waste generation, Composition and management in the Amhara National Regional State, Ethiopia 2015. Dissertation for the Degree of Doctor of Philosophy (PhD) Addis   Ababa   University,   Ethiopia   July,   2015 (https://www.academia.edu/39270146/HEALTHCARE_WASTE_IN_ETHIOPIA_A_STUDY_OF_W ASTE_GENERATION_COMPOSITION_AND_MANAGEMENT_IN_THE_AMHARA_NATIONA L_REGIONAL_STATE_ETHIOPIA_ESUBALEW_TESFAHUN_Dissertation_for_the_Degree_of_D octor_of_Philosophy).

[61]   Oke IA. Management of immunization solid wastes in Kano State, Nigeria. Waste Manag 2008; 28(12): 2512-21.
[http://dx.doi.org/10.1016/j.wasman.2007.11.008] [PMID: 18191394]

[62]   Olorunfemi FB. Living with waste: Major sources of worries and concerns about landfills in lagos metropolis. Nigeria Ethiopian J Environ Stud Manag 2009; 2(2)

[63]   Ketlogetswe C, Oladiran MT, Foster J. Improved combustion processes in medical wastes incinerators for rural applications. Afr J Sci Technol 2004; 5(1): 67-72.

[64] Botswana Environmental and Climate Change Analysis 29 May, 2008. This Environmental and Climate Change Analysis was written at the request of Sida INEC, Stockholm (att: Rolf Folkesson) by Gunilla Ölund Wingqvist and Emelie Dahlberg at Sida Helpdesk for Environmental Economics, University of Gothenburg as part of Sida-EEU"s institutional collaboration on environmental economics and strategic environmental assessment (https://www.efdinitiative.org/sites/default/files/0531sidaeeurep0709.pdfoptimized.pdf).

[65] Ikome P, Mochungong K. Environmental Exposure and Public Health Impacts of Poor Clinical Waste Treatment and Disposal In Cameroon Thesis .2011.(https://europepmc.org/article/med/28299096).

[66] Mbuligwe SE, Kaseva ME. Pollution and self-cleansing of an urban river in a developing country: a case study in Dar es Salaam, Tanzania. Environ Manage 2005; 36(2): 328-42.
[http://dx.doi.org/10.1007/s00267-003-0068-4] [PMID: 16025200]

[67] Gyambrah R. MSc Final Thesis Msc. Disaster and Risk Mgt- Hazardous Waste 2017 (https://www.slideshare.net/RansfordGyambrahMISQ/ransford-gyambrah-fin-l-masters-thesis-2017-open-university-of-malaysia).

[68] KYERE VINCENT NARTEY Environmental and Health Impacts of Informal E-waste Recycling in Agbogbloshie. Recommendations for Sustainable Management. Accra, Ghana 2016.

[69] Sibanda T, Selvarajan R, Tekere M. Urban effluent discharges as causes of public and environmental health concerns in South Africa's aquatic milieu. Environ Sci Pollut Res Int 2015; 22(23): 18301-17.
[http://dx.doi.org/10.1007/s11356-015-5416-4] [PMID: 26408112]

[70] ACAP. Project "Reduction/Elimination of dioxin and furan emissions in the Russian Federation with Focus on the Arctic and Northern Regions Impacting the Arctic" Phase II report 2003.

[71] ACAP. 2005.Assessment of mercury releases into the environment from the territory of the Russian Federation. The Federal Service for Environmental, Technological and Nuclear Supervision in cooperation with the Danish Environmental Protection Agency, Copenhagen http://www.zeromercury.org/library/Reports%20General/0502%20Dk%20report%20on%20Hg%20releases%20in%20

[72] Goss PE, Strasser-Weippl K, Lee-Bychkovsky BL, *et al.* Challenges to effective cancer control in China, India, and Russia. Lancet Oncol 2014; 15(5): 489-538.
[http://dx.doi.org/10.1016/S1470-2045(14)70029-4] [PMID: 24731404]

[73] Franchini M, Rial M, Buiatti E, Bianchi F. Health effects of exposure to waste incinerator emissions:a review of epidemiological studies. Ann Ist Super Sanita 2004; 40(1): 101-15.
[PMID: 15269458]

[74] Vaverková MD Landfill Impacts on the Environment—Review Geosciences. 2019; 9(10): 431.

[75] Saunders P. Population health and waste management: scientific data and policy options. Report of a WHO workshop. Rome, Italy 2007.

[76] Editorial. A toxic legacy Illegal dumping of toxic waste in the Italian Campania has been blamed for high rates of ill health in the region. The suspected link needs to be investigated using the most modern methods. 24 APRIL 2014; 508: 431.

[77] Senior K, Mazza A. Italian "Triangle of death" linked to waste crisis. Lancet Oncol 2004; 5(9): 525-7.
[http://dx.doi.org/10.1016/S1470-2045(04)01561-X] [PMID: 15384216]

[78] Forte IM, Costa A, Iannuzzi CA, *et al.* Blood screening for heavy metals and organic pollutants in cancer patients exposed to toxic waste in southern Italy: A pilot study. J Cell Physiol 2019.
[http://dx.doi.org/10.1002/jcp.29399] [PMID: 31838757]

[79] Richardson SD. Water analysis: emerging contaminants and current issues. Anal Chem 2009; 81(12): 4645-77.
[http://dx.doi.org/10.1021/ac9008012] [PMID: 19456142]

[80] Nolan LA, Nolan JM, Shofer FS, Rodway NV, Emmett EA. The relationship between birth weight,

gestational age and perfluorooctanoic acid (PFOA)-contaminated public drinking water. Reprod Toxicol 2009; 27(3-4): 231-8.
[http://dx.doi.org/10.1016/j.reprotox.2008.11.001] [PMID: 19049861]

[81]   Birnbaum LS, Staskal DF. Brominated flame retardants: cause for concern? Environ Health Perspect 2004; 112(1): 9-17.
[http://dx.doi.org/10.1289/ehp.6559] [PMID: 14698924]

[82]   Kolpin DW, Furlong ET, Meyer MT, *et al.* Pharmaceuticals, hormones, and other organic wastewater contaminants in U.S. streams, 1999-2000: a national reconnaissance. Environ Sci Technol 2002; 36(6): 1202-11.
[http://dx.doi.org/10.1021/es011055j] [PMID: 11944670]

[83]   Schultz MM, Higgins CP, Huset CA, Luthy RG, Barofsky DF, Field JA. Fluorochemical mass flows in a municipal wastewater treatment facility. Environ Sci Technol 2006; 40(23): 7350-7.
[http://dx.doi.org/10.1021/es061025m] [PMID: 17180988]

[84]   Mitch WA, Sharp JO, Trussell RR, Valentine RL, Alvarez-Cohen L. Sedlak DL. N-nitrosodimethylamine (NDMA) as a drinking water contaminant: areview. Environ Eng Sci 2003; 20: 389-404.
[http://dx.doi.org/10.1089/109287503768335896]

[85]   IARC, 2014. World Cancer Report, 2014. WHO, Lyon. (https://publications.iarc.fr/Non-Serie--Publications/World-Cancer-Reports/World-Cancer-Report-2014)

[86]   World Health Organization. Population health and waste management: scientific data and policy options Report of a WHO workshop Rome, Italy, 29–30 March 2007. Copenhagen: WHO Regional Office for Europe 2007.

CHAPTER 2

Hazardous Waste and its Management

Gabriella Marfe[1,*], Stefania Perna[1] and **Carla Di Stefano[2]**

[1] *Department of Scienze e Tecnologie Ambientali, Biologiche e Farmaceutiche, University of Campania "Luigi Vanvitelli," via Vivaldi 43, Caserta 81100, Italy*

[2] *Department of Hematology, "Tor Vergata" University, Viale Oxford 81, 00133Rome, Italy*

Abstract: Hazardous Waste Management (HWM) in developed and developing countries faces several challenges. Such waste requires special attention throughout all its stages of management to avoid the contamination of soil, groundwater and surface water. In this regard, the growing amount of waste generated has made it increasingly important to develop strategies to manage waste safely through accurate disposal practices. Reducing, reusing and recycling waste is the most effective way to create a safe waste management in a suitable environmentally friendly way. Therefore, every country should adopt a policy to reduce the polluted impacts on environmental and global health.

Keywords: Hazardous Waste Management, Health risks, Incinerators, Landfills.

INTRODUCTION

Disposal of hazardous waste is becoming an environmental and public health issue. The hazardous wastes are generated by both small companies (dry cleaners, auto repair shops, hospitals, exterminators, and photo processing centers) and big companies (including chemical manufacturers, electroplating companies, and petroleum refineries). In addition, there are hazardous household wastes as well such as batteries, gasoline, antifreeze products, oil-based paints and thinners and household cleaning products. The reduction of hazardous waste through its minimization or the use of replaceable materials plays a key role in waste management Unfortunately, such waste is disposed of unsafely, thereby poisoning the environment and causing damage to the health of both humans and animals. In an important study, Nema and Gupta [1] reported *"the possibilities to ensure safe,*

* **Corresponding author Marfe Gabriella:** Department of Scienze e Tecnologie Ambientali, Biologiche e Farmaceutiche, University of Campania "Luigi Vanvitelli," *via* Vivaldi 43, Caserta 81100, Italy; Tel: +39 0823 275104, Fax: +39 0823274813; E-mail: gabriellamarfe@gmail.com

efficient and cost-effective collection, transportation, treatment and disposal of wastes".

Today, the treatment facilities for this waste are incinerators and landfills, that are undesirable because the populations do not want to live near them. Another problem of hazardous waste management is associated with the generation of different residues that should be treated and disposed of in a proper manner. The volume or mass reduction obtained after incineration is higher when compared with the volume or mass reduction obtained after chemical disinfection. Additional problems are the transportation cost of these residues, identification of appropriate facilities and location of the disposal facilities. Furthermore, the right technologies to recycle these residues are an essential issue to decrease their amount. After the treatment process, both non-recycled and recycled residues produced by hazardous wastes should be routed to the right disposal facility (Fig. 1).

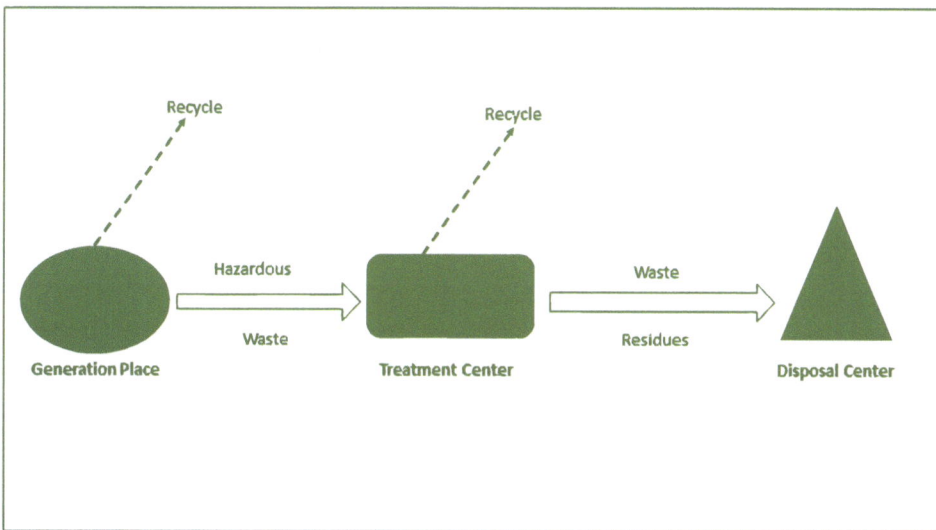

Fig. (1). Hazardous waste management problem.

HAZARDOUS WASTE

Hazardous Waste Definition

In 1970, in a study, conducted in the USA, the term "hazardous waste" was used for the first time [2]. It is very difficult to define hazardous waste in a single manner [3]. Its definition should include (a) waste characteristics; (b) effects on humans and environment; (c) source of waste; and (d) cause of generation. The United Nations Environmental Programme [4] (as cited in LaGrega, Buckingham and Evans, 2001) defines hazardous waste as follows:

Hazardous wastes mean wastes (solids, sludges, liquids, and containerized gases) other than radioactive (and infectious) wastes which, by reason of their chemical activity or toxic, explosive, corrosive, or other characteristics, cause danger or likely will cause danger to health or the environment, whether alone or when coming into contact with other waste (p. 2).

In the US, The Resource Conservation and Recovery Act [5] (RCRA, as cited in Wentz, 1989) legislation of 197 in the USA, has considered hazardous waste as:

... a solid waste, or combination of solid wastes, which because of its quantity, concentration, or physical, chemical, or infectious characteristics may (1) cause, or significantly contribute to, an increase in mortality or an increase in serious irreversible or incapacitating reversible illness or (2) pose a substantial present or potential hazard to human health or the environment when properly treated, stored, transported or disposed of or otherwise managed (p. 1).

Furthermore, the US Environmental Protection Agency (US EPA) has defined hazardous waste considering the nature, composition, inherent characteristics and the source [5] (as cited in Wentz, 1989, p. 1). Hazardous waste:

1. has characteristics of ignitability, corrosivity, reactivity, and/or toxicity;

2. is a non-specific source waste (generic waste from industrial processes);

3. is a specific commercial chemical product or intermediate;

4. is a mixture containing a listed hazardous waste;

5. is a substance that is not excluded from regulation under the Resource, Conservation and Recovery Act, Subtitle C (as reported by Origins of the Convention, by UNEP, n.d., Retrieved April 13, 2002, from HYPERLINK "http://www.basel.int/pub/basics.html"\l "intro%3B"\h http://www.basel.int/pub/basics.html#intro; Hazardous waste management (p. 2) by LaGrega, Buckingham, and Evans, 2001, Singapore: McGraw-Hill.; and Hazardous waste management (p. 1), by Wentz, 1989, New York: McGraw-Hill).

Furthermore, there are two systems for hazardous wastes classification [5, 6]:

Exclusive List System

This system includes groups of non-hazardous wastes and it is accepted by a few developed countries.

Inclusive List System

The list of this system includes only hazardous chemicals and wastes and it is recognized by different countries (such as Belgium, Denmark, France, Federal Republic of Germany, The Netherlands, India, Sweden, United Kingdom, and United State of America).

Three different methods are used to consider the waste as hazardous in this system:

a) **"General listing approach"**, based on an assumption of hazard or non-hazard without test;

b) **"Listing approach"**, prior listing as hazardous wastes based on the profile of the potential health effects;

c) **"Listing by characteristics"**, identification of characteristics by testing and declaration of hazard.

These definitions can vary because of different contexts and/or functions and responsibilities in a range of organizations.

Characteristic Hazardous Waste

Wastes are defined as hazardous when they present four different characteristics as described above: ignitability, corrosivity, reactivity, and toxicity (Article 3 of Chapter 11 of the hazardous waste regulations -Sections 66261.21 to 66261.24).

A. Ignitable wastes are dangerous because they can cause a fire under particular conditions, be subjected to spontaneous combustion, or have a flash point less than 60°C (140°F). This characteristic is defined in section 6626121 of the hazardous waste regulations. In order to define the ignitability of wastes three different tests are used: the Pensky-Marteus Closed-Cup Method for Determining Ignitability, the Seta flash Closed-Cup for Determining Ignitability, and the Ignitability of Solids.

B. Corrosive wastes are acids or bases materials, or they produce acidic or alkaline solutions when react with other compounds. A liquid waste is considered corrosive when it is able to corrode metal containers, such as storage tanks, drums, and barrels. The characteristic of corrosivity is defined in section 6626122 of the hazardous waste regulations. In order to define corrosive wastes two tests are used: pH Electronic Measurement and Corrosivity towards Steel.

C. Reactive wastes are unstable under normal conditions. They can cause explosions or release toxic fumes, gases, or vapors when heated, compressed, or mixed with water. The characteristic of reactivity is defined in section 66261.23 of the hazardous waste regulations. In order to evaluate their reactivity, the narrative criteria reported in the hazardous waste regulations are used.

D. Toxic wastes are harmful or fatal when ingested or absorbed (*e.g.* wastes. containing mercury, lead, DDT, PCBs, *etc*). The disposal of toxic wastes can be dangerous because their constituents might leak from the waste and pollute groundwater. The characteristic of toxicity is defined in section 66261.24 of the hazardous waste regulations. Furthermore, it contains eight subsections, to identify the toxicity of wastes:

• **TCLP:** Toxic as defined through the application of a laboratory test procedure called the Toxicity Characteristic leaching Procedure (TCLP - U S EPA Test Method 131 1). The TCLP identifies wastes (as hazardous) that may leach hazardous concentrations of toxic substances into the environment. The result of the TCLP test is compared to 'The Regulatory Level (RL) in the table in subsection 66261.240) (1) of the hazardous waste regulations. This criterion does not apply to wastes that are excluded from regulation under the Resource Conservation and Recovery Act.

• **Total and WET:** Toxic as defined through application of laboratory test procedures called the "total digestion" and the "Waste Extraction Test" (commonly called the "WET"). The results of each of these laboratory tests are compared to their respective regulatory limits, the Total Threshold Limit Concentrations (TTLCs) and the Soluble Threshold Limit Concentrations (STLCs), which appear in subsection 66261.24(a) (2) of the hazardous waste regulations.

• **Acute Oral Toxicity:** Toxic because the waste is an acutely toxic substance or contains an acutely toxic substance., if ingested. As stated in subsection 66261.24(a)(3), a waste is identified as being toxic if it has an acute oral LD5O less than 2,500 mg/kg A calculated oral LD5O may be used.

• **Acute Dermal Toxicity:** Toxic because the waste is either an acutely toxic substance or contains an acutely toxic substance if dermal exposure occurs. As stated in subsection 66261.24(a)(4), a waste is identified as being toxic if it has a dermal LD5O less than 493190 mg/kg. A calculated dermal LD5O may be used.

• **Acute Inhalation Toxicity:** Toxic because the waste either is an acutely toxic substance or contains an acutely toxic substances if inhaled As staled in subsection 66261.24 (a)(5), a waste is identified as being toxic if it has a dermal LC50 less than 10.000 U S. EPA Test Method, SW-846 Methods: 381Os Headspace (formerly Method 5020) may be used to "lest out" (for volatile organic substances).

• **Acute Aquatic Toxicity:** Such waste is toxic to fish. A waste is aquatically toxic if it produces a LC5O less than 500 mg/L when tested using the "Static Acule Bioassay Procedures for Hazardous Waste Samples". Such test procedure is available at:

http:/www.dtsc.ca.gov/HardousWaste/upload/HWMP_bioassay_report.pdf.

• **Carcinogenicity:** Toxic because it contains one or more carcinogenic substances. As stated in subsection 66261,24 (a)(7), a waste is identified as being toxic if it contains any of the specified carcinogens al a concentration of greater than or equal to 0 001 percent by weight.

• **Experience or Testing:** Pursuant to subsection 6626224(a)(8), a waste may be toxic (and therefore, a hazardous. waste) even if it is not identified as toxic by any of the seven criteria above. At the present time, only wastes containing ethylene glycol spent antifreeze solutions) have been identified as toxic by this subsection.

Hazardous Waste Generation and Impacts

Hazardous waste can be classified considering the generated quantity 1) Very Small Quantity Generators (VSQGs); 2) Small Quantity Generators (SQGs); 3) Large Quantity Generators (LQGs) [7]. Very Small Quantity Generators (VSQGs) generate 100 kilograms or less per month of hazardous waste or one kilogram or less per month of acutely hazardous waste. Small Quantity Generators (SQGs) generate more than 100 kilograms, but less than 1,000 kilograms of hazardous waste per month, while Large Quantity Generators (LQGs) generate 1,000 kilograms per month or more of hazardous waste or more than one kilogram per month of acutely hazardous waste. For example, a big industry generates over 1000 kg per month of hazardous, while vehicle maintenance, equipment repair, construction, printing, photography, laboratories, schools, laundries generate less than 1000 kg per month of hazardous waste [7]. Naturally, it is important the careful management during transporting, operating, storing, and the prevention of illegal dumping in public areas as in order to avoid negative effects as follows [8]

(Greenberg and Anderson, 1984, p. 51, as cited in Salcedo, Cross, & Chrismon, 1989):

1. Effects on people: chronic illness, chronic disability, minor and temporary illness.

2. Effects on the environment: ecosystem elimination, elimination or reduction of some species, reduction of biomass, loss of water sources.

3. Social effects: negative effects on communities (on disadvantaged people, future generation).

4. Economic effects: devaluation of property, pollution of the land and local taxpayers pay for cleanup, security, and other maintenance of site.

Hazardous Waste Management

Hazardous waste management includes pollution prevention through recycling and reuse as reduction, recycling, treatment, storage and disposal in order to eliminate the generation of hazardous waste. As shown in Fig. (2), the pollution prevention techniques as suggested by the EPA's policy statement are as follows:

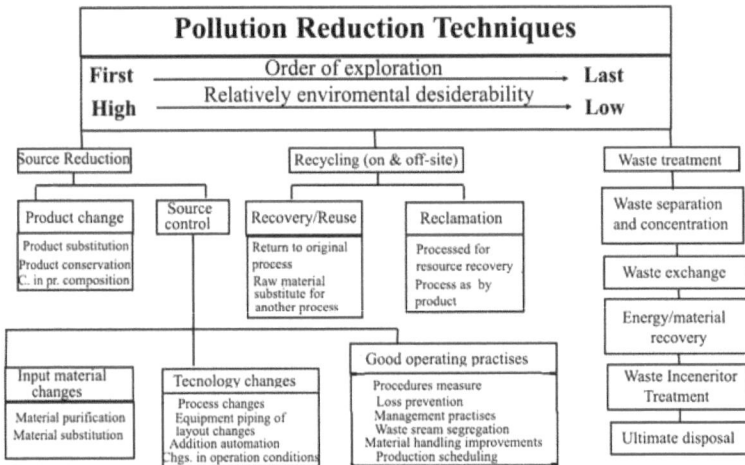

Fig. (2). Pollution prevention techniques. (adapted by Hazardous waste management (p. 373), by LaGrega, Michael D., Buckingham, Phillip L., and Evans, Jeffrey C. 1994, McGraw-Hill, New York).

1. Source reduction;

2. Recycling: when it is not possible to recycle the residues of hazardous waste it

is necessary to treat them in a safety manner;

3. The hazardous waste or their residues should be treated in an environmentally safe manner when the treatment (source reducing and recycling) are not possible;

4. The hazardous waste is should be disposed of in an environmentally proper manner when it is impossible to treat them.

Hazardous waste facilities include multiple technologies to treat different types of hazardous waste and their residues [5]. The major types of facilities of such waste consist of recovery/recycling, treatment, storage and disposal. The recovery or recycling facilities are able to reduce and conserve quantities of materials, but they are also costly, and for this reason, still unpopular [5]. The treatment facilities include different biology process such as physical, chemical, thermal, or biological treatment, since these wastes can request different technologies to reduce their toxicity and their volume [5, 9]. The disposal facilities include places where it is possible to collocate hazardous wastes after their treatment [10]. For this reason, there are different type of these facilities such as open dumps, surface impoundments, sanitary landfills, "secure" chemical landfills, underground injection wells, land treatment facilities, salt domes, salt bed formations, underground mines, underground caves, ocean dumping, incineration, and immobilization [8, 11, 12]. Moreover, generators are called "on-site management", when they are able to manage hazardous waste through one process or several processes on their own land and, while other generators are called "off-site management" when they are obliged to send their hazardous wastes to the specialized operational firms providing management in treatment and disposal. The facilities operations system can be split out five subsystems [10]: 1) pre-shipment waste analysis: it is necessary to analyze the produced hazardous wastes by scientific technologies before treatment, store and disposal; 2) waste receiving: generally, the transport of hazardous wastes have to include proper documents about their treatment and disposal in the facilities schedule in order to avoid illegal dumping. Waste is controlled, weighed and divided into several samples to analyze different parameters. Next, waste is dumped at the unloading area where it is stored and kept safe manner before treatment and /or disposal. The hazardous wastes must be singled out and in order to undergo the proper treatment (mixing, blending, and repackaging methods) and to store them into a suitable status. Other types of waste undergo treatment processes before being mixed and treated with biological technologies. These operations should be carefully monitored along each step by instruments, direct human observation, and chemical analysis to avoid mistakes along this assembly line. For example, some residuals derived by treatment processes such as gaseous emissions and wastewater effluents have to dispose of in safety manner. Furthermore, special measures such as security, inspection maintenance, incident prevention, emergency planning, closure plans are necessary to finish the facility's operation in a safe manner for future management of this waste.

LANDFILL DISPOSAL OF HAZARDOUS WASTE

The hazardous wastes can be located in landfill disposal after source reduction, recycling, and treatment [12]. This disposal facility is the most widespread method of eliminating hazardous waste for different reasons: a) the cost of this method is cheap; b) the procedure is unsophisticated; c) the environmental legislation, during the 1970s and early 1980s, did not permitted the release of hazardous waste into the atmosphere or water ways, and d) this disposal was believed to be safe and proper [8, 12].

Landfill Disposal

Today, landfill disposal plays still an important role among the different technologies to dispose of one part of hazardous waste. Although there are multiple hazardous waste treatments such as waste minimization, recycle and reduction, it is very difficult to eliminate wastes totally and for this reason, some residues, derived from hazardous waste treatment technology are disposed of in landfill [4].

Methods of Landfill Disposal

There are two types of landfills: sanitary and secure landfills. The first one is a well-engineered and well-controlled open dump, suitable only for solid "non-hazardous waste". Such waste is located and compacted into a layer a few feet thick, then covered and compacted again at least every day. Generally, the liners are natural and synthetic, and these landfills have a leachate detection and collection system, leachate cap, and groundwater in order to monitor odor, fires, water contamination, and wind-blown wastes. However, there are a lot of problems with these landfills because of leaks, bed smells, generate gas (vinyl chloride and methane) and nuisance [13]. Moreover, this kind of landfills can be harmful because improper management of ignitable, reactive, or incompatible wastes can cause fires, explosions, toxic fumes, groundwater contaminations. The main cause of these phenomena is the mixing of household and industrial wastes. Several researchers have shown that the chemical reactions can happen among unknown and unpredictable kinds of wastes from those households and industries. Furthermore, the biological treatment improvement of incompatible waste and the innovation of landfill design can decrease their contamination [8]. The secure landfill disposal represents the ultimate placement of hazardous waste to reduce the pollution to humans and environment through air, water, and/or land. Nevertheless, these landfills can be secure when hazardous waste management has been performed in a proper manner from collection, prevention, minimization, storage, and treatment [12]. Consequently, the design and the construction of these landfills are carefully planned and engineered.

Selection of the Site

Generally, the choice of a suitable place to build a landfill needs to consider the geological, hydrological characteristics of the land. In this context, it is essential to select the sites by avoiding floodplains, wetlands, earthquake-prone areas, historical places, habitats for endangered species, breeding or stopping off areas for migratory birds, as well as prime agricultural land and proximity to human habitats [4, 8]. The details of the landfill site selection criteria shown in Table **1**.

The landfill disposal should be well prepared to avoid the worst risk in environment and human health. For this issue, it is very fundamental to consider these objectives [5]: (Wentz, 1989, p. 317):

1. Type and volume of hazardous and non-hazardous wastes to be disposed of.

2. Life expectancy of the landfill during its active operating period.

3. Topography and soil characteristics at the site and in its vicinity.

4. Climatic conditions throughout the year.

5. Surface water and groundwater in the vicinity.

6. Collection and treatment of surface run-off.

7. Soil cover requirements for individual containment cells.

8. Anticipated quality and volume of leachate.

9. Selection of leachate collection and treatment systems.

10. Monitoring of groundwater and surface water during operation and beyond.

11. Selection of flexible membrane and other impermeable liners.

12. Closure and post-closure plans.

13. Alternatives used during the post-closure period.

14. Effect on human health and the environment.

The hazardous wastes should undergo the right treatment technologies before disposing of a landfill. First of all, these wastes are tracked during the transportation from their point of generation to their ultimate disposal site. Then, they are classified before being placed into a landfill. Here, the waste layers are covered and compacted by thickness of soil (30, 48 cm) to decrease bad smell, airborne transport of contaminants, the potential for direct contact, and then, are monitored until to the post-closure period [10].

Generators of hazardous waste facilities are responsible for safely closing and post-closure of the facilities, as well as any unexpected occurrences. Additionally, they monitor air emissions, leachate, and landfill wells to avoid the pollution of the neighborhood environment from hazardous waste.

There are different objectives to have a secure landfill disposal facility:

1. The control of the top to reduce air emission, run-off and infiltration of precipitation.

2. The control of the bottom to decrease the collection of leachate run-off control and to minimize contaminant transport through the bottom.

3. The installation of adequate leachate collection, treatment systems and monitoring wells.

Naturally, it is important to know soil and rock characteristics, groundwater levels, flood levels, access to transportation and acceptability in order to select the right site for the construction of a secure landfill (see Table 1).

Table 1. Landfill Site Selection Criteria.

Engineering	
Physical site	Should be large enough to accommodate waste for the life of production facility.
Proximity	Locate as close as possible to production facility to minimize handling and reduce transport cost. Locate away from water supply (suggested minimum 500 feet) and property line (suggested minimum 200 feet).
Access	Should be all-weather, have adequate width and loan capacity, with minimum traffic congestion. Easy access to highways and railways transport.
Topography	Should minimize earth-moving; take advantage of natural conditions. Avoid natural depression and valleys where water contamination is likely (suggested site slope of less than 5%).

(Table 1) cont.....

Engineering	
Geology	Avoid areas with earthquakes, slides, faults, underlying mines, sinkholes and solution cavities.
Hydrology	Areas with low rainfall and high evapotranspiration and not affected by ideal water movements and seasonal highwater table.
Soils	Should have natural clay liner or clay available for liner, and final cover material available; stable soil/rock structure. Avoid sites with thin soil above groundwater, highly permeable soil above shallow groundwater and soils with extreme erosion potential.
Drainage	Areas where surface drainage exists and can be easily controlled.
Environmental	
Surface water	Locate outside 100-year floodplain. No direct contact with navigable water. Avoid wetlands.
Groundwater	No contact with groundwater. Base of fill must be above high groundwater table. Avoid sole-source aquifer and areas of ground water recharge.
Temperature	Not within area of recurring temperature inversion.
Air/wind direction	Areas where prevailing wind will carry-away any emission and odour from populated areas or ecologically sensitive areas.
Terrestrial and aquatic ecology	Avoid unique habitat areas (important to propagation of rare and endangered species) and wetlands. Avoid national parks, forests, flora, fauna reserves, and coastal areas.
Public Health	Areas where construction and operation will not adversely affect public health.
Aesthetic	Sites where minimum visual impact is created owing to construction and operation; sites should be designed considering surrounding landscape.
Noise	Minimize truck traffic and equipment operation noise.
Land use	Avoid populated areas and areas of conflicting land use such as parks, scenic area, labour intensive industrial sectors, recreational reserves, camp sites, sporting reserves, intensive agricultural area, area zoned for future urban development, and so on.
Cultural Resources	Avoid areas of unique archaeological, historical and paleontological interest.
Infrastructure	
Power and water	Areas with easy access to adequate power and water-supply.
Sewer:	Site near interceptor sewer or wastewater treatment plants.
Legal/regulatory	Consider national, regional and local requirements for permits.
Public/political:	Gain local acceptance from elected officials and local interest groups.

Note. From Batstone, Roger; Smith, James E. Jr.; Wilson, David; WHO; UNEP.1989. The safe disposal of hazardous wastes: the special needs and problems of developing countries: Volume 1 (English). World Bank technical paper; no. WTP 93. Washington, D.C: World Bank Group. http://documents.worldbank.org/curated/en/695131468764392542/Volume-1.

This kind of landfill should be constructed with double composite liners and a leachate collection system above and between the liners. Moreover, they have a leak detection system that it is able to remove any leakage between the liners at the earliest practicable time (Fig. **3**). Landfill liners should be built with particular materials that have different chemical properties such as strength and thickness to avoid failures caused by pressure gradients. In addition, it is essential to prevent physical contact with waste or leachate, considering that they are vulnerable to climatic conditions, the stress of both installation and daily operation the stress of installation and daily operation. Selection of a particular liner depends on [12].

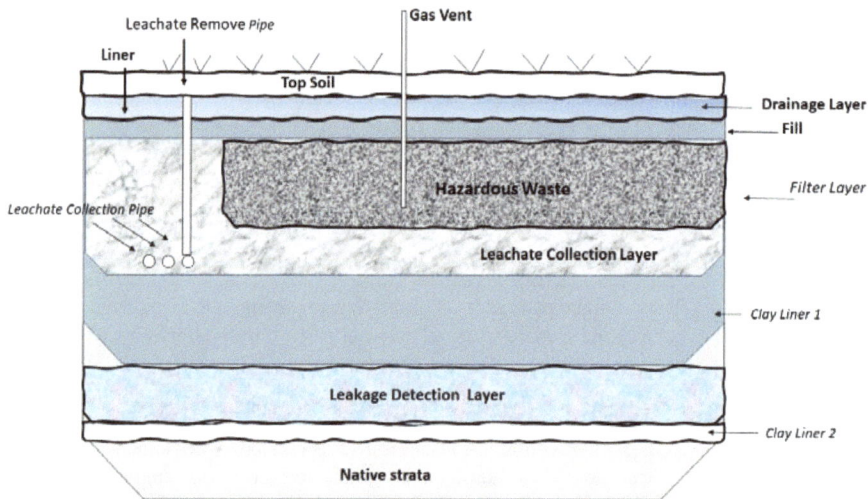

Fig. (3). Cross section of hazardous waste landfill disposal (adapted from US EPA, 1989- Final Covers on Hazardous Waste Landfills and Surface Impoundments EPA 530-SW-89-047 **PB89-233480** 1989-Fig. **7**).

1. Effectiveness: liner type and waste type.

2. Cost: both installation and acquisition.

3. Installation time.

4. Durability.

Problems with Landfill Disposal

Landfills disposal management can create several risks to human health and environment. The weaknesses of choosing landfill technology, synthetic and landfill liners can have the following disadvantages:

1. Their expected lifetime cannot be established, but their effectiveness can last for several years.

2. Waste disposal operations can break the liner, provoking leachate seepage.

3. Changes in hydraulic conductivity of the underlining or surrounding soil increase the level of the groundwater, which generates upward pressure on the liner.

4. When the liner is in place and waste is deposited, it can be difficult to detect the liner failure and for this reason, liner it is not possible to readily repair.

However, inappropriate practice managements and failures of landfill disposal technology can a great impact on human health and environment. Many studies have shown that landfill disposal of hazardous waste can result in pollution of air, water and land [11, 14].

INCINERATION OF HAZARDOUS WASTE

Methods of Incineration

Modern waste incinerators can destroy many kinds of hazardous wastes, for example:

1. Biologically hazardous wastes.

2. Wastes which are resistant to biodegradation and persistent in the environment.

3. Liquid wastes which are highly volatile and, therefore, can be easily dispersed.

4. Liquid wastes which have flash point below 40°C.

5. Wastes which cannot be disposed of in a secured landfill site.

6. Wastes containing organically bounded halogens, lead, mercury, cadmium, zinc, nitrogen, phosphorus or sulfur.

Incineration or thermal oxidation is a process in which hazardous wastes are burnt using the oxygen present in the air and then transformed it into gases and incombustible solid residue. Consequently, such process leads a progressive reduction of waste that is sent to landfill [12]. This process depends on different physical parameters such as temperature, time, and turbulence (mixing). To completely burn and eliminate wastes, the temperature should be sufficiently high, and the time should be very long in order to reduce also air emission residues. However, this practice creates large amounts of heat. Thus, the precise temperature and optimum time for an incineration operation have to be considered. In addition, it is also important to control the mixing of the wastes and oxygen through the degree of turbulence in the incinerator to obtain ta good destruction of the waste. Furthermore, it is possible to burn waste alongside traditional fossil fuels like coal in facilities such as cement kilns, coal-fired power plants and industrial boilers through the co-incineration process [12]. Moreover, the air emissions derived from incinerator (in the form of odors, particulates, and hazardous gaseous substances) need to be controlled because they can contain hydrogen chloride, carbon monoxide, sulfur dioxide, nitrogen oxides, heavy metals, and other ash particulate matter which in turn can cause pollution through wastewater and runoff stormwater contamination. Besides solid residues derived from incineration should be disposed of in a secure landfill.

Site Selection Models

Another difficult task is site selection of an incinerator for two important reasons:1) it should be considered for its natural features and the land use to avoid possible danger to public health and the environment, 2) it should be accepted by local communities. LaGrega, Buckingham, and Evans (2001) [10] suggested that possible hazardous waste facility sites had to choose after being undergone to specific criteria: this is the "phase approach". The literature analysis specifies three sequential periods (technical, political and psychological), for the search of sites selection. Details of these site selection periods are given below:

• The First Period is characterized by Technical Approach" that can be divided in the "closed siting approach" or "decide-announce-defend (DAD)" [15] (Khun, Ballard, 1998, p. 535); the"technical decision" [16] (DeSario, Langton, 1987b, p. 211) the "pre-emptive approach" [17] (Portney, 1991, p. 9) and the "top-down approach" [18] (Rabe, 1992, p. 119). The technical approach screens possible site through computerized geographic information systems (GIS) [10], and the mathematic model calculation [19], rather than social and political aspects. Examples of a closed siting approach exist in British Columbia and Ontario, Canada [15]; examples of the pre-emptive approach exist in Illinois, Maryland and Florida, US [17]. The Closed siting approach bases on the top-down process considering geological and biophysical constraints for site selection. The approach often fails because of low attention to social and political considerations [15].

• The Second Period is characterized by Social Approach. Previously, some hazardous waste sites were chosen, considering only high technical engineering selection knowledge and less public participation in decision making. For example, some hazardous waste facilities sites have been constructed within minority and disadvantaged communities, which then had negative results [20]. In this case, the population could not make decisions in the process of site selection [21, 22]. This is the beginning of the "social approach". In the "social approach", "open siting approach", or "establish-criteria-consult-filter-decide (ECFD)" [15] (Khun, Ballard, 1998, p. 536), "value decision" [16] (DeSario, Langton, 1987b, p. 211) or the "state authority with local input approach" [17] (Portney, 1991, p. 9), are considered the social aspects rather than on the technical and engineering aspects [16]. The open siting approach considers only the willing communities to open a hazardous waste site close to them. The interested communities can be responsible for providing a potential site that meets the basic environmental criteria, as well as making sure the majority of the local residents support the project by a referendum. Practical examples of this approach exist in Alberta and Manitoba, Canada. Another example of hazardous waste facility site selection is that of a private sector project: willing landowners or communities are involved in the process. They can propose their own land such as a hazardous waste facility site even if it has imperfect characteristics for site selection. It means that private developers can invest a lot of money to develop it into ideal land for a hazardous waste facility [10]. In the Second Period it is possible considered three Mixed Approaches (a) the "multi-attribute utility analysis technique (MUA)" [23, 24] (Keeney, Raiffa, 1976, cited in Merkhofer, Conway, Anderson, 1997, p. 832)—an analytic approach for making logical decisions when there are multiple objectives, (b) the "mixed decision" [16] (De Sario and Langton, 1987b, p. 211), and (c) the "local involvement and influence approach" [17] (Portney, 1991, p. 10). After the screening of site selection, the representatives from the stakeholders make a decision. In this case, both the developer and the public have substantial influence over the siting process. Moreover, the MUA approach requires only representative stakeholders without public involvement [16]. Before making decisions, people attend technical meetings, interpret scientific presentations and controversies, and logically resolve conflicting social desires and capabilities [24].

• The Third period is characterized by Risk Substitution Strategy. In this case, the public acceptance of hazardous waste facilities site can depend on risk perception. It means that the population will accept the facility if the existing risk is reduced or the existing risk is higher than the risk from this facility, such as a nuclear factory. This approach includes four steps of decision making: 1) the sites are screened for both technical and social issues; 2) the possible sites are narrowed down after the screening. 3) the "trade-off" between the existing facilities and the new facilities is assessed. It is necessary to make an in-depth environmental assessment between the existing risk and the new risk. However, this approach has two limitations; first, the public will not accept any trade-off if there is no existing significant risk perception; and second, it is not clear what role the media would play in this risk substitution strategy.

Site Selection

The successful location of a hazardous waste facility depends on the perception of risk and acceptance by the population, including both technical and political issues. The first period (as described above) is characterized by a technical approach to choose adequate sites through computerized geographic information systems (GIS) and mathematic model calculation. The second period (as described above) is represented by increased participation, awareness and acceptance by the population. In the end, the third period is characterized by Risk Substitution Strategy: the population can accept hazardous waste facilities site when it is possible to reduce risks for human health and environment. In this regard, numerous studies show the optimum model of hazardous waste facility site selection [10, 16 - 24].

STATUS OF HAZARDOUS WASTE MANAGEMENT

The hazardous wastes are normally generated by the chemical and allied products industry. New data show that the generation of hazardous waste is increasing at a rate of 2–5% per year. The estimated quantity of industrial hazardous waste indicate that the total production in the Organization for Economic Co-operation and Development (OECD) countries was some 300 million of tons in 1985 out of which the United States accounted for some 268 million of tons, European OECD countries 24 million of tons, and the balance of 8 million of tons was contributed by the Pacific OECD countries. In most developing countries, there is a lack of data inventory on hazardous waste management. UNEP-WHO ad-hoc Working Group of Experts on Environmentally-sound Management of Hazardous Wastes concluded in 1984 that "it is safe to say that virtually all developing countries have yet to develop a comprehensive hazardous waste management scheme, as compared to those already established in most industrialized countries". Today, even after 18 years, very little progress has been made [5]. Even for an advanced country like the United States, as late as mid-1979, its Council on Environmental Quality reported that "existing estimates on the extent of the problem vary so widely that they are only marginally helpful". In the same year, US EPA (the U.S. Environmental Protection Agency.) estimated that only about 10% of hazardous wastes were being disposed of in a manner that was likely to comply with the regulations. A similar situation exists now in most of the developing countries [5]. In these countries, it has been observed that due to the non-availability of any organized waste disposal system, hazardous wastes are being handled unscientifically. The indiscriminate disposal of hazardous waste mixed with municipal solid waste (MSW) in open lands (such as low-lying area or along road sides) is a common practice. It is surprising to note that the developing nations rather than learning from the mistakes of developed countries are repeating the same actions, so aggravating the environmental problem. Several studies are carried out to better understand the possible health effects related to residential proximity to landfills or incinerators. In two reviews, the authors did not find the elevated cancer incidence in people living close to landfill, while they reported an increased incidence in low birth weight and congenital malformation [25, 26]. Another review pointed out the health effects of municipal solid waste on individuals that worked in the collection, transport, transfer and management of solid waste [27]. Some recent studies, carried out in Italy and Spain, reported contrasting data: the Italian study [28] observed an increase of respiratory disease in people living close to municipal landfill in Rome, while there was no evidence about this relationship in the Spain study [29, 30]. Specifically, the Italian study was in accordance with other papers [31 - 33]. In Finland, Pukkala *et al.* [34] studied cancer and chronic disease incidence in people living in houses built on a precedent dumping site that contained industrial and household wastes. The authors found an elevated number of male cancers (especially for pancreas and of the skin) after adjusting for age and sex. Other authors considered [35] the cancer risks in people living within 2 km of about 9,565 (from a total of 19,196) landfills, located in Great Britain from 1982 to 1997. They did not find high risk of cancers (such as bladder cancer, brain cancer, hepato-biliary cancers or leukaemia), after adjusting for age, sex, calendar year and deprivation. Using data of municipal solid waste incinerators from a previous study [36], Knox [37] reported about 22,458 cancer deaths occurring before the 16th birthday in Great Britain between 1953 and 1980. He analyzed 9,224 cancer cases, comparing the distance from suspected sources to the birth addresses and to the death addresses. He reported increased cancer/leukaemia rate in children who moved away from birthplaces close to municipal incinerators. Another study in France [38] found an increased incidence of non-Hodgkin's lymphoma (NHL) and soft tissue sarcoma in people living around a municipal solid waste incinerator with high dioxin emissions. Floret *et al.* [39] conducted a population-based case-control study in the same area. The authors compared the risk cancer in people living close to area with the high dioxin levels respect to those living close to area with the low dioxin concentrations. They found that the risk of developing cancers (such as non-Hodgkin lymphoma) was higher in people living in arear with elevated dioxin levels. Based on these findings, a nationwide study was conducted in the vicinity of some municipal solid waste incinerators with high dioxin emission levels in France [40]. The data found a correlation between non-Hodgkin's lymphoma (NHL) incidence and exposure to dioxins emitted by municipal incinerators. Different studies addressed the use of the human biomonitoring for exposure assessment of pollution due to incinerators. Overall, typical biomarkers have smaller variance ratios than other measurements (such as air measurement) in environmental settings [41]. For example, different papers reported the association between high polycyclic aromatic hydrocarbon (PAH) levels in human urine and increased dioxin emissions from incinerators [42]. As already stated in the previous article [43], suggested a promising way forward in the use of human biomonitoring to improve the assessment of exposure to environmental stressors. Furthermore, Pastor *et al.* [44] conducted a study in the dog living close to a municipal solid waste incinerator with high dioxin emissions. The authors found a positive association between canine NHL and the distribution of waste incinerators. Another study found no association between exposure to dioxin and breast cancer in women younger or older than 60 years, respectively, living near a French municipal solid waste incinerator with high exposure to dioxin [45].On the contrary, two previous studies suggested a higher overall risk of breast cancer among women living in Chapaevsk (Russia) and Seveso (Italy) [46, 47]. In Italy, Biggeri *et al.* [48] carried out a case-control study in Trieste in order to investigate the association between several sources of environmental pollution and lung cancer. Using spatial models and distance from the sources, the authors studied 755 cases of lung cancer. Specifically, they observed a statistically high correlation between lung cancer and distance from the polluted sources after adjustment for age, smoking habits, the likelihood of exposure, occupational carcinogens, and levels of air particulate. An excess risk of lung cancer was also found in females living in two areas of the province of La Spezia (Italy) exposed to environmental pollution emitted by multiple sources, including an industrial waste incinerator [49]. In another case-control study, Comba *et al.* [50] found an increased risk of soft tissue sarcomas in people living within two km of an industrial waste incinerator in the city of Mantua. The same risk was found by Zambon *et al.* [51] in the province of Venice. The authors carried out case-control study in residents close to 10 municipal solid waste incinerators with high dioxin emissions. They observed a statistically significant association between the risk of sarcoma and the levels of dioxin-like substances. The results were more significant for women than for men. Another study was conducted in four French administrative departments and the authors used dispersion modelling, advanced geographical information system (GIS) and statistical techniques to estimate the association between the risk of cancer and the exposure to incinerators. They pointed out a high risk of cancer morbidity for residents around municipal solid waste incinerators [52]. High lung cancer mortality was found among males living close to Italian National Priority Contaminated Sites (with industrial waste landfills or illegal dumps) [53] and among residents living near incinerators and landfills of hazardous waste in Spain [54]. However, the conducted studies are insufficient to support the links between the elevated levels of dioxin emitted by incinerators and the high incidence of lung cancer [55]. Another residential cohort study, conducted in a contaminated site located in the suburb of Rome (Italy), found that Hydrogen sulfide (H_2S) exposure from the landfill was related to higher risk of mortality from laryngeal cancer and bladder cancer in women, as well as hospitalizations for cardiorespiratory diseases [56]. The first biomonitoring case study was conducted in volunteers that lived close to incinerator of Mataró in Spain [57, 58]. In according with age, sex and distance to the incinerator exposure, the authors measured the PCDD/Fs and PCBs levels in whole blood sample pools. The concentrations of PCDD/Fs and PCBs were found higher in women respect with men, and in older age group in comparison to the younger ones. Other studies were carried out on the population living near two Portuguese incinerators (one in Lisbon and one in Madeira Island). The results detected that the dioxin levels in blood and human milk were higher in population living in Lisbon than in Madeira [59 - 66]. Another study considered the incinerators (49 in Italy, 2 in Slovakia and 11 in England) and landfills (619 in Italy, 121 in Slovakia and 232 in England) in three European countries. Such study was conducted on population living within 3 km from incinerators and 2 km from landfilld. Air pollution dispersion modelling for particulate matter (PM10) and nitrogen dioxide (NO_2) was used to evaluate the excess risk estimates. The authors measured attributable cancer incidence and years of life lost (YoLL) for incinerators and attributable cases of congenital anomalies and low birth weight infants for landfills. They found that 1,000,000, 16,000, and 1,200,000 subjects lived close to incinerators in Italy, Slovakia and England, respectively. The additional contribution to NO_2 levels within a 3 km radius was 0.23 (Italy), 0.15 Slovakia), and 0.14µg/m^3 (England), while PM10 values were very low Considering that the incinerators continue to work until 2020, the authors suggested that the annual number of cancer cases due to exposure between 2001-2020 will reach 11 (Italy), 0 Slovakia), and 7 (England) in 2020 and then will decline to 0 in the three countries in 2050. Additionally, they supposed that by 2050, the attributable impact on the 2001 cohort of residents will be 3,621 (Italy), 37 (Slovakia) and 3,966 (England) YoLL. Moreover, they estimated that 1,350,000 (Italy), 329,000 (Slovakia) and 1,425,000 (England) subjects living close to landfills. In this case, they believed that the annual additional cases of congenital anomalies up to 2030 will be approximately 2 (Italy), 2 (Slovakia) and 3 (England), whereas additional low-birth weight newborns will be 42 in Italy, 13 in Slovakia, and 59 in England [67]. Another report showed that children living close the incinerator in Sint-Niklaas (Belgium) developed blood and glandular cancers after five years, while adult people had a five-fold risk of developing different kinds of cancer over 20 years [68]. Furthermore, from the year 1961 up to 1969, orofacial defects incidence was increased near to the incinerator of Zeeburg. Amsterdam [69]. In the EUROHAZCON study, Dolk *et al.* found that residents living within 3 km from a landfill site were associated with a significantly increased risk of non-chromosomal congenital anomaly (adjusted odds ratio [(OR) 1.33 95% CI 1.11–1.59]. Odds ratios for specific congenital anomalies were also significantly raised for residents within 3 km of a site (OR for neural-tube defects: OR for malformations of the cardiac septa: 1.49 [1.09–2.04], for anomalies of great arteries and veins the OR were: 1.81 [1.02–3.20] [70, 71].

In the USA, after the Love Canal incident and the creation of Superfund National Priorities List (NPL), the first epidemiologic studies of NPL site found that white males in the county hosting the site had a significantly elevated odds of developing bladder cancer (OR = 1.7, p<0.025) [72]. The authors measured exposure at the county level, using national averages for comparison, and multiple outcomes were evaluated, so the elevated risk might have been due to multiple comparisons [72]. A study found that residents living close to Superfund site in Texas reported more neurological symptoms compared to low-exposure populations in Texas [73], and that incidence rates for multiple cancers were elevated in the vicinity of a Department of Defense Superfund site in Massachusetts [74]. Studies have also examined the degree to which contaminants could migrate from Superfund sites into the surrounding ecosystems and communities. Three bark samples within 10 km of a NPL site in Michigan showed 10-100 fold increases in dichlorodiphenyltrichloroethane (DDT), hexabromobenzene, and polybrominated biphenyls compared to sites >10 km distant. Passive sampling devices that mimicked the way living organisms accumulate lipophilic contaminants were deployed near a NPL site in Portland, Oregon, and contaminant levels that would result in an excess cancer risk greater than the EPA limit of 1x10-6 were found [75]. The addition of geospatial information and tools in public health research have increased precision for examining spatial patterns within data, understanding relationships between outcomes and environmental variables, and inferring exposure patterns [76]. When precise address information is available for cases, geospatial analysis can provide sharp, precise boundaries of a cluster or area of exceedance to most efficiently deploy public health resources [77]. As evaluated areas get smaller (*e.g.* county, census tract, census block, geospatial coordinates), there is less variability in exposures, and ecological fallacy becomes less likely [78]. A geographic distribution analysis showed that blood levels of dieldrin (an organochlorine insecticide) increased by 1.6 ng/g for each one mile of closer residential proximity to a Superfund site in Maryland [79]. Another study of a NPL site contaminated with polychlorinated biphenyls (PCBs) found that residential proximity to the site was not a significant factor in cord serum PCB levels, but being born before or during dredging activities to remove PCBs from the site was significant [80]. Geospatial analysis has also been used to identify clusters of childhood cancer near NPL sites in Dade County, Florida [81], and very-low birth weight near multiple NPL sites in Harris County, Texas [82]. In another study, the authors found an association between people living close to hazardous waste sites containing benzene and hospitalization discharge rates for persons having hematologic cancers in New York State except for New York City for the years 1993 to 2008. Moreover, they observed a 15% increase in the rate of hospitalization for chronic lymphatic leukemia (CLL), a 22% increase in the rate of discharges for total leukemia and a 17% increase in the rate of discharges for total lymphoma [83]. To assess the relationship between bladder cancer risk and the presence of toxic release inventory (TRI) sites in Utah's census tracts, some authors performed a geographical case-control study on the stable areas comparing the proportion of areas with TRI sites between high risk areas (93 areas) and the others (390 areas) for both genders. Of the 22 census tracts with TRI sites over the 10 years, 21 showed a stable spatial odd ratios (OR). The odds ratio (OR) for presence of TRI sites was 2.73 (95% CI = 1.05-6.69, both genders). When restricted to males only, the distribution of the TRI sites were similar and the OR was 2.96 (95% CI = 1.14-7.17). This suggested a positive association between presence of TRI sites and higher risk of bladder cancer for both genders and for males only [84]. The report entitled Health Alert: Disease Clusters Spotlight the Need to Protect People from Toxic Chemicals showed disease clusters occurred in USA [85]. Researchers of the National Resources Defense Council (NRDC) and National Disease Clusters Alliance (NDCA) focused on only thirteen states, Texas, California, Michigan, North Carolina, Pennsylvania, Florida, Ohio, Delaware, Louisiana, Montana, Tennessee, Missouri, and Arkansas, chosen on the basis of occurrence of known clusters in the state, geographic diversity, or community concerns about a disease cluster in their area. For example, birth defects in Kettleman City, California, including twenty babies born over less than two years with birth defects, and four children born with birth defects so severe that they have since died, in this town of only 1,500 people, that live close to hazardous waste disposal facility (3.5 miles southwest of town). A cluster of breast cancer was identified in an urban census tract at the Agricultural Street Landfill Superfund Site in New Orleans (Orleans Parish). According to Agency for Toxic Substances and Disease Registry (ATSDR), the site and the neighborhood were contaminated with metals, polyaromatic hydrocarbons (PAHs), volatile organic compounds, and pesticides. There was evidence that PAHs could increase the risk of developing breast cancer. Over the period of 1986 through 1987, a cluster of neuroblastoma was identified by researchers at Louisiana State University Medical School in Amelia, St. Mary Paris. In 1994, ATSDR concluded that there was evidence to suggest that adverse health outcomes in the community could be related to environmental exposures. However, there were insufficient data to link hazardous waste incinerator at Marine Shale Processor (MSP) to adverse health outcomes in the community. In

2006, MSP and its owner paid the state government a settlement of $7 million for the closure and remediation of the site. The health assessment investigated the higher rates of health problems, including leukemia, birth defects and childhood total cancer and the high incidence of respiratory problems in Ellis County (Midlothian Texas), when compared to the rest of the state. Researchers at Southwest Texas State University found a cluster of liver cancer deaths in Bexar County (San Antonio California) and its adjacent counties using statewide cancer mortality data from 1990 through 1997. About 14 zip codes in San Antonio encompassed a plume of polluted groundwater linked to Kelly Air Force Base. Local groups alleged that the groundwater was polluted with waste containing benzene, perchloroethylene, and trichloroethylene, all known carcinogens. Finally, in 2015 the U.S. Environmental Protection Agency (EPA) released a new vapor intrusion guidance document, in which was specified two conditions: (1) proof of a vapor intrusion pathway; and (2) evidence that human health risks exceed established thresholds (for example, one excess cancer among 10,000 exposed people). However, the guidance lacks details on methods for demonstrating these conditions In a recent review, Jill Johnston and Jacqueline MacDonald Gibson have recommended methods to improve monitoring, modeling and integrating information to support remediation decisions at vapor intrusion sites, building on EPA's new guidance document [86].

Different reports highlighted that countries of Central and Eastern Europe (such as Czech Republic, Hungary, Slovenia, Belarus, Latvia, Ukraine, Estonia, Poland, Lithuania, Croatia, Moldova, Georgia, the Kyrgyz Republic, Russia, Tajikistan and Armenia) had a hazardous waste management systems very similar to those of developing countries. In another study, Navarro *et al.* demonstrated the potential of novel real-time GPS in birds for detecting illegal urban dumps [87]. This research investigated the spatial movements of yellow-legged gulls (Larus michahellis), a model species of scavenging gull keen on human organic waste. Adult birds carried solar-powered GPS trackers, and the combined weight of the GPS devices plus harness was less than 2.5% of the body mass of the bird. GPS data were analyzed into the central database and available in the Virtual Lab for Bird Movement Modelling - UvA-BiTS. This study showed that gulls equipped with real-time GPS could give immediate information about the presence and location of illegal urban waste dumps. Sanneh and others described the situation about waste management in Banjul (the Gambian capital West Africa). In particular, the dumpsite was located in a densely populated area, where the residents were affected by the smoke from burning debris and the bed smell of decomposing waste. Furthermore, the nuisances worst during rainy time since the area becomes infested with flies and insects [88]. Other studies showed an exponential increase in the volume of waste generation in different countries such as Nigeria [89, 90], Botswana [91] and Cameroon [92] because of rapid industrialization and intense population growth [93, 94]. For example, in Phnom Penh, capital city of Cambodia, in 2015, about 685,000 tons of municipal solid waste were generated in unsafe manner because of facilities lack for the waste management [95]. In Thailand, more than 60% of the SMW was disposed of at open dumping sites [96]. In East Timor, a study observed that population did not sort their waste, because they had not enough education and awareness in terms of treating the waste properly [97]. Another problem is the expansion of the population in the cities that results in an elevated waste generation due to the level of daily consumption [98]. For instance, one study, conducted in Sri Lanka by the Asian Institute of Technology, reported that the population were increased by 32.5% annually in association with an increase of waste generation. In this regard, in this nation, the implementation of sustainable waste management is a difficult task [99]. In Kenya, a study by Ziraba *et al.* [100] pointed out the association between poor waste management and adverse health outcomes. A research study by Al Khatib *et al.* observed that open dumps was still utilized to dispose of waste in seven Palestinian districts. In recent years, many developing countries struggle with the amount of plastic waste [101]. While the United States, Japan and many European countries generate significant amounts of plastic waste, they are also relatively good at managing it. About half of all of the plastic waste that ends up in the oceans comes from just five countries: China, Indonesia, the Philippines, Vietnam and Thailand [102]. Around 40% of plastic waste originated from packaging products as a result of unsustainable consumption patterns [103]. In Maputo, administrative center of Mozambique, the SMW are disposed of in the official landfill, where open fires and autoignition of the waste happened very often [104]. Moreover, the landfill at Mathkal (Kolkata, India) is worsening water quality because of presence of several substances such as Cd and Ni (detectable in leachate) and some metals Pb, Cd, Cr and Ni [105]. In Chennai city, the capital of Tamil Nadu, India, elevated concentrations of heavy metals were found in the soil samples at various depths (ranges from 3.78 mg kg^{-1} to 0.59 mg kg^{-1} at a depth of 2.5 to 5.5 m) [106]. At the end, in Nonthaburi dumpsite, Thailand, the heavy metals such as chrome, cadmium, lead, nickel and mercury, were detected in boreholes and runoff in high concentration (10 times above the limits introduced by the World Health Organization (WHO) for drinking waste) [107]. In an impressive study conducted in three Asian countries, researchers reported health problem in the population who was exposed to open dumping or burning waste for the release of dangerous compounds such as dioxins. Chatham-Stephens and his collaborators have developed a disability-adjusted life year (DALY)-based estimate of the disease burden attributable to toxic waste sites in the low- and middle-income countries (LMICs): India, Indonesia, and the Philippines. The authors calculated that about 8,629,750 persons could be exposed to industrial pollutants derived from 373 toxic waste sites in these countries in 2010. They observed 54,432 DALYs attributable to lead exposure in India (n = 24), 78,982 DALYs in Indonesia (n = 28) and 394,084 DALYs in the Philippines (n = 27) [108]. In another study, the authors used the DALY and the exposure assessment method developed by Chatham-Stephens *et al.* They identified 679 contaminated hazardous waste sites in seven Asian countries. Among them, 169 sites were polluted by lead and the authors calculated that about 245,949 (0–4 years old) children were exposed to lead. Such levels of exposure might be sufficient to generate acute and chronic adverse effects, such as a decrease in Intelligence Quotient (IQ). In this study, they found 45,321 DALYs attributable to lead-contaminated sites in Mexico (n = 62), furthermore they estimated 0.31 DALYs(3,1) per person for exposure to lead in Argentina, Mexico and Uruguay (mean population at risk of exposure per site = 2455). In this analysis, the calculated burden of disease due to exposure to lead was very similar to that calculated for Parkinson's disease and bladder cancer in these countries [109]. In Mexico, the waste management system is not safe since approximately 24% of the total waste is burned by households in rural areas, while in Brazil, the waste is managed in improper manner: 41.3% of the municipal solid waste was inappropriately disposed of uncontrolled landfills or open dumps [110]. Furthermore, Brazilian towns produce high proportions of organic waste and there is no system for the separation of organic and non-organic waste in houses [111, 112]. Another study, conducted in Copey (Colombia), found an increased number of cancers and birth defect cases within the community living close to an illegal dumpsite. Over 200 tons of hazardous chemicals were found at the site, including DDT3 [113]. The hazardous waste management business in Colombia is now emerging under two somewhat antagonist circumstances. From one side, there is a new comprehensive regulatory framework under which hazardous waste dumping or mismanagement would be heavily penalized. But, from another side, the country does not have (such as in Bogotá) the proper infrastructures to cover a growing demand for hazardous waste management, so the government neither can fully enforce these regulations nor fully comply with international treaties on hazardous waste control [114].

In 2012, the World Bank reported that the cities generated 1.3 billion tons of municipal solid waste at global level. The same organization has estimated a production of 2.2 billion tons per year by 2025, considering both population and urbanization increase. In this context, it seems crucial waste reduction following zero waste strategy (composting, recycling and reuse). There are different lab initiatives engaged in development of the future zero waste cities around the world (such as the Adelaide Living Laboratory" (ALL), the New Zealand based Zero Waste Academy (ZWA-LL), and the "ECO LivingLab@Chamusa" in Portugal). These structures try to support interdisciplinary research between the university and the host city. Furthermore, there are several Zero Waste initiatives around the world. Capannori was the first one, declaring a Zero Waste 2020 goal in 2007. This is a small town of the province of Lucca (Tuscany Italy) with 45,850 population. In 1999, the municipality started a system of door-to-door collection of sources, separated discards. A study carried out by La Sapienza University in Rome compared door-to-door collection in three communities in Italy (Capannori, Rome, Salerno). The authors reported that in 99% of inhabitants of Capannori separated their waste compared to other two communities. By 2010, 82% of municipal waste was separated, just 18% of residual waste is disposed of landfill (https://zerowasteeurope.eu/2013/09/the-story-of-capannori-a-zero-waste-champion/). Italy is also one of the countries committed to zero waste and 232 cities are working towards this goal. Argentona (Catalunya town) was the first city to adopt zero waste system in Spain. From 2004, such city doubled its recycling rates. Another Spanish province Gipuzkoa led to the adoption of a door-to-door waste collection service, reducing the amount of waste going to landfills by 80 percent (https://www.foeeurope.org/sites/default/files/news/case_study_argentona_english_final_ok_small.pdf). The Slovenian capital, Ljubljana and three other municipalities, Vrhnika, Borovnica and Log Dragomer have adopted Zero waste system. Vrhnika reached 82% in separate collection of municipal solid waste, while the other cities collected and recycled over 76% of their waste. In 2018, Sălacea, a small town in Romania with a population of 3,000, started waste collection system, separating organic waste, glass and plastic packaging. In the first month, 66% of waste was recycled compared to only 3% previously. In the United States, most of the places adopting zero waste goals are at the municipal county or regional level. According to the Zero Waste International Alliance [115], California is the only state that has officially adopted zero waste goal. San Francisco has reduced its waste to landfill by 78% through incentives and extensive public outreach. As for the rest of the world, according to ZWIA, 25 international countries, regions, and cities have indicated adherence to zero waste goals. Australia and South Africa are two nations that have made zero waste commitments. Several case studies of zero waste innovation point out that significant progress can be obtained in environmental action. Moreover, it can be affirmed that such a strategy could improve its goals in accordance with UN's Sustainable Development Goals (SDGs).

CONCLUSION

It is necessary to create new rules for hazardous wastes management:

1. Reduction of hazardous wastes generation;

2. Re-use and re-processing of wastes;

3. Waste exchange: (it means that hazardous wastes from one industry can be transferred to another industry, which can use them as raw materials);

4. Separation of hazardous and non-hazardous wastes;

5. The transformation of hazardous wastes into non-hazardous;

6. Disposal of hazardous wastes in a secure landfill: after minimization, reuse/recycle, exchange, and proper treatment of its residues;

7. Roles and responsibilities of agencies involved in hazardous waste management;

8. Capacity Building and Awareness Creation for an effective and adequate waste management strategy.

CONSENT FOR PUBLICATION

Not applicable.

CONFLICT OF INTEREST

The authors confirm that the contents of this chapter have no conflict of interest.

ACKNOWLEDGEMENTS

Declare none.

REFERENCES

[1]   Nema AK, Gupta SK. Optimization of regional hazardous waste management systems: an improved formulation. Waste Manag 1999; 19: 441-51.
        [http://dx.doi.org/10.1016/S0956-053X(99)00241-X]

[2]   Sinha RK. Hazardous Wastes: The Technological Wastes of the Modern Civilisation: An Overview. Griffith University: Brisbane, Australia 2002b.

[3]   Blackman WC. Basic Hazardous Waste Management. Boca Raton, Florida: Lewis 2001.

[4]   LaGrega MD. Buckingham PL, Evans JC. Hazardous Waste Management. New York: McGraw-Hill 1994.

[5]   Wentz CA. Hazardous Waste Management. New York: McGraw-Hill 1989.

[6]   Saxena AK, Gupta Y. Saxena AK, Gupta Y. Environmentally sound Management of Hazardous Wastes Hazardous Waste Management – Environmentally Sound Management of Hazardous Wastes Encyclopedia of Life Support Systems (EOLSS) (https://www.eolss.net/Sample-Chapters/C09/E1--8-20-00.pdf).

[7]   Polprasert C, Liyanage LRJ. Hazardous waste generation and processing. Resour Conserv Recycling 1996; 16: 213-26.
        [http://dx.doi.org/10.1016/0921-3449(95)00058-5]

[8]   Salcedo RN, Cross FL, Chrismon RL. Environmental Impacts of Hazardous Waste Treatment Storage and Disposal Facilities. Pennsylvania: Technomic 1989.

[9]   Saxena SC, Jotshi CK. Management and combustion of hazardous waste. Pror Energy Combust Sci 1996; 22: 401-25.
[http://dx.doi.org/10.1016/S0360-1285(96)00007-X]

[10]   LaGrega MD, Buckingham PL. Evans JC. Hazardous Waste Management. Singapore: McGraw-Hill 2001.

[11]   Blackman WC. Blackman WC Jr. Basic Hazardous Waste Management. 3rd ed. Boca Raton, Florida: Lewis 2001.

[12]   Visvanathan C. Hazardous waste disposal. Resour Conserv Recycling 1996; 6: 201-12.
[http://dx.doi.org/10.1016/0921-3449(95)00057-7]

[13]   Liu DHF, Lipták BG. Hazardous Waste and Solid Waste. London: Lewis 2000.

[14]   Sinha RK. Hazardous Wastes and Substances-A Time Bomb: Management Strategies and Technologies-Prevention the Best Remedy. Griffith University: Brisbane, Australia 2002c.

[15]   Kuhn RG, Ballard KR. Canadian Innovations in siting hazardous waste management facilities. Environ Manage 1998; 22(4): 533-45.
[http://dx.doi.org/10.1007/s002679900126] [PMID: 9582390]

[16]   DeSario J, Langton S. Toward a Metapolicy for Social Planning. New York: Greenwood 1978b.

[17]   Portney KE. Siting Hazardous Waste Treatment Facilities: The NIMBY Syndrome. New York: Auburn House 1992.

[18]   Rabe BG. When siting works, Canada-style. J Health Polit Policy Law 1992; 17(1): 119-42.
[http://dx.doi.org/10.1215/03616878-17-1-119] [PMID: 1619246]

[19]   Giannikos I. A multiobjective programming model for locating treatment sites and routing hazardous wastes. Eur J Oper Res 1998; 104: 333-42.
[http://dx.doi.org/10.1016/S0377-2217(97)00188-4]

[20]   Oakes JM, Anderton DL. Anderson AB. A longitudinal analysis of environmental equity in communities with hazardous waste facilities. Soc Sci Res 1996; 25: 125-48.
[http://dx.doi.org/10.1006/ssre.1996.0006]

[21]   Wright SA. The Not-In- My- Backyard Syndrome: A research proposal for assessing public resistance. J Hazard Mater 1989; 22: 258.
[http://dx.doi.org/10.1016/0304-3894(89)85060-5]

[22]   The hazardous waste facility siting controversy: the Massachusetts experience. Am J Law Med 1987; 12(1): 131-56.
[PMID: 3661576]

[23]   Keeney L, Raiffa H. Decisions with Multiple Objectives: Preferences and Value Trade-Offs-R. New York: Wiley 1976.

[24]   Merkhofer MW, Conway R, Anderson RG. Profile: Multiattribute utility analysis as a framework for public participation in siting a hazardous waste management facility. Environ Manage 1997; 21(6): 831-9.
[http://dx.doi.org/10.1007/s002679900070] [PMID: 9336482]

[25]   Rushton L. Health hazards and waste management. Br Med Bull 2003; 68: 183-97.
[http://dx.doi.org/10.1093/bmb/ldg034] [PMID: 14757717]

[26]   Linzalone N, Bianchi F. Inceneritori: non solo diossine e metalli pesanti, anche polveri fini e ultrafini. Epidemiol Prev 2007; 31(1): 62-6. [Incinerators: not only dioxins and heavy metals, also fine and ultrafine particles]. [in Italian].
[PMID: 17591406]

[27]   Perez HR, Frank AL. Zimmerman NJ. Health effects associated with organic dust exposure during the handling of municipal solid waste. Indoor Built Environ 2006; 15: 207-12.

[28]   Mataloni F, Badaloni C, Golini MN. Morbidity and mortality of people who live close to municipal waste landfills: a multisite cohort study. Int J Epidemiol 2016; 45(3): 806-15.
[http://dx.doi.org/10.1093/ije/dyw052] [PMID: 27222499]

[29]   Gouveia N, Prado RR. Health risks in areas close to urban solid waste landfill sites. Rev Saude Publica 2010; 44(5): 859-66. a
[http://dx.doi.org/10.1590/S0034-89102010005000029] [PMID: 20882262]

[30]   Gouveia N, Prado RR. [Spatial analysis of the health risks associated with solid waste incineration: a preliminary analysis]. Rev Bras Epidemiol 2010; 13(1): 3-10. b [Portuguese.].
[http://dx.doi.org/10.1590/S1415-790X2010000100001] [PMID: 20683550]

[31]   Heaney CD, Wing S, Campbell RL, *et al.* Relation between malodor, ambient hydrogen sulfide, and health in a community bordering a landfill. Environ Res 2011; 111(6): 847-52.
[http://dx.doi.org/10.1016/j.envres.2011.05.021] [PMID: 21679938]

[32]   Corrêa CR, Abrahão CE, Carpintero MdoC, Anaruma Filho F. Landfills as risk factors for respiratory disease in children. J Pediatr (Rio J) 2011; 87(4): 319-24.
[PMID: 21590001]

[33]   Mattiello A, Chiodini P, Bianco E, *et al.* Health effects associated with the disposal of solid waste in landfills and incinerators in populations living in surrounding areas: a systematic review. Int J Public Health 2013; 58(5): 725-35.
[http://dx.doi.org/10.1007/s00038-013-0496-8] [PMID: 23887611]

[34]   Pukkala E, Pönkä A. Increased incidence of cancer and asthma in houses built on a former dump area. Environ Health Perspect 2001; 109(11): 1121-5.
[http://dx.doi.org/10.1289/ehp.011091121] [PMID: 11712996]

[35]   Jarup L, Briggs D, de Hoogh C, *et al.* Cancer risks in populations living near landfill sites in Great Britain. Br J Cancer 2002; 86(11): 1732-6.
[http://dx.doi.org/10.1038/sj.bjc.6600311] [PMID: 12087458]

[36]   Elliott P, Shaddick G, Kleinschmidt I, *et al.* Cancer incidence near municipal solid waste incinerators in Great Britain. Br J Cancer 1996; 73(5): 702-10.
[http://dx.doi.org/10.1038/bjc.1996.122] [PMID: 8605111]

[37]   Knox E. Childhood cancers, birthplaces, incinerators and landfill sites. Int J Epidemiol 2000; 29(3): 391-7.
[http://dx.doi.org/10.1093/ije/29.3.391] [PMID: 10869308]

[38]   Viel JF, Arveux P, Baverel J, Cahn JY. Soft-tissue sarcoma and non-Hodgkin's lymphoma clusters around a municipal solid waste incinerator with high dioxin emission levels. Am J Epidemiol 2000; 152(1): 13-9.
[http://dx.doi.org/10.1093/aje/152.1.13] [PMID: 10901325]

[39]   Floret N, Mauny F, Challier B, Arveux P, Cahn JY, Viel JF. Dioxin emissions from a solid waste incinerator and risk of non-Hodgkin lymphoma. Epidemiology 2003; 14(4): 392-8.
[http://dx.doi.org/10.1097/01.ede.0000072107.90304.01] [PMID: 12843761]

[40]   Viel JF, Daniau C, Goria S, *et al.* Risk for non Hodgkin's lymphoma in the vicinity of French municipal solid waste incinerators. Environ Health 2008; 7: 51.
[http://dx.doi.org/10.1186/1476-069X-7-51] [PMID: 18959776]

[41]   Lin YS, Kupper LL, Rappaport SM. Air samples *versus* biomarkers for epidemiology. Occup Environ Med 2005; 62(11): 750-60.
[http://dx.doi.org/10.1136/oem.2004.013102] [PMID: 16234400]

[42]   Cordioli M, Ranzi A, De Leo GA, Lauriola P. A review of exposure assessment methods in

epidemiological studies on incinerators. J Environ Public Health 2013; 2013129470
[http://dx.doi.org/10.1155/2013/129470] [PMID: 23840228]

[43]   WHO Regional Office for Europe. 2007.www.euro.who.int/document/e90174.pdf

[44]   Pastor M, Chalvet-Monfray K, Marchal T, *et al.* Genetic and environmental risk indicators in canine
       non-Hodgkin's lymphomas: breed associations and geographic distribution of 608 cases diagnosed
       throughout France over 1 year. J Vet Intern Med 2009; 23(2): 301-10.
       [http://dx.doi.org/10.1111/j.1939-1676.2008.0255.x] [PMID: 19192140]

[45]   Viel JF, Clément MC, Hägi M, Grandjean S, Challier B, Danzon A. Dioxin emissions from a
       municipal solid waste incinerator and risk of invasive breast cancer: a population-based case-control
       study with GIS-derived exposure. Int J Health Geogr 2008; 7: 4.
       [http://dx.doi.org/10.1186/1476-072X-7-4] [PMID: 18226215]

[46]   Revich B, Aksel E, Ushakova T, *et al.* Dioxin exposure and public health in Chapaevsk, Russia.
       Chemosphere 2001; 43(4-7): 951-66.
       [http://dx.doi.org/10.1016/S0045-6535(00)00456-2] [PMID: 11372889]

[47]   Warner M, Eskenazi B, Mocarelli P, *et al.* Serum dioxin concentrations and breast cancer risk in the
       Seveso Women's Health Study. Environ Health Perspect 2002; 110(7): 625-8.
       [http://dx.doi.org/10.1289/ehp.02110625] [PMID: 12117637]

[48]   Biggeri A, Barbone F, Lagazio C, Bovenzi M, Stanta G. Air pollution and lung cancer in Trieste, Italy:
       spatial analysis of risk as a function of distance from sources. Environ Health Perspect 1996; 104(7):
       750-4.
       [http://dx.doi.org/10.1289/ehp.96104750] [PMID: 8841761]

[49]   Parodi S, Baldi R, Benco C, *et al.* Lung cancer mortality in a district of La Spezia (Italy) exposed to air
       pollution from industrial plants. Tumori 2004; 90(2): 181-5.
       [http://dx.doi.org/10.1177/030089160409000204] [PMID: 15237579]

[50]   Comba P, Ascoli V, Belli S, *et al.* Risk of soft tissue sarcomas and residence in the neighborhood of an
       incinerator of industrial wastes. Occup Environ Med 2003; 60: 650-83.
       [http://dx.doi.org/10.1136/oem.60.9.680]

[51]   Zambon P, Ricci P, Bovo E, *et al.* Sarcoma risk and dioxin emissions from incinerators and industrial
       plants: a population-based case-control study (Italy). Environ Health 2007; 6: 19.
       [http://dx.doi.org/10.1186/1476-069X-6-19] [PMID: 17634118]

[52]   Goria S, Daniau C, de Crouy-Chanel P, *et al.* Risk of cancer in the vicinity of municipal solid waste
       incinerators: importance of using a flexible modelling strategy. Int J Health Geogr 2009; 28:8:31
       [http://dx.doi.org/10.1186/1476-072X-8-31]

[53]   Fazzo L, Minichilli F, Pirastu R, *et al.* A meta-analysis of mortality data in Italian contaminated sites
       with industrial waste landfills or illegal dumps. Ann Ist Super Sanita 2014; 50(3): 278-85.
       [PMID: 25292275]

[54]   García-Pérez J, Fernández-Navarro P, Castelló A, *et al.* Cancer mortality in towns in the vicinity of
       incinerators and installations for the recovery or disposal of hazardous waste. Environ Int 2013; 51:
       31-44.
       [http://dx.doi.org/10.1016/j.envint.2012.10.003] [PMID: 23160082]

[55]   Porta D, Milani S, Lazzarino AI, Perucci CA, Forastiere F. Systematic review of epidemiological
       studies on health effects associated with management of solid waste. Environ Health 2009; 8: 60.
       [http://dx.doi.org/10.1186/1476-069X-8-60] [PMID: 20030820]

[56]   Ancona C, Badaloni C, Mataloni F, *et al.* Mortality and morbidity in a population exposed to multiple
       sources of air pollution: A retrospective cohort study using air dispersion models. Environ Res 2015;
       137: 467-74.
       [http://dx.doi.org/10.1016/j.envres.2014.10.036] [PMID: 25701728]

[57]   Gonzalez CA, Kogevinas M, Gadea E, *et al.* Biomonitoring study of people living near or working at a

municipal solid-waste incinerator before and after two years of operation. Arch Environ Health 2000; 55(4): 259-67.https://hero.epa.gov/hero/index.cfm/reference/details/reference_id/2138342
[http://dx.doi.org/10.1080/00039890009603416] [PMID: 11005431]

[58]   González CA, Kogevinas M, Gadea E, Pera G, Päpke O. Increase of dioxin blood levels over the last 4 years in the general population in Spain. Epidemiology 2001; 12(3): 365.https://insights.ovid.com/crossref?an=00001648-200105000-00020
[http://dx.doi.org/10.1097/00001648-200105000-00020] [PMID: 11338318]

[59]   Reis MF, Miguel JP, Sampaio C, Melim JM, Aguiar P. First results from dioxins and dioxin-like compounds in the population from Madeira Island, Portugal. Part 1 - biomonitoring in blood of the general population living near to a solid waste incinerator. Organohalogen Compd 2004; 66: 2702-8.

[60]   Reis MF, Sampaio C, Aguiar P, Matos P, Paepke P, Pereira MJ. Dioxin contamination status in people living near a Portuguese Municipal Solid Waste. Organohalogen Compd 2005; 67: 1552-5.

[61]   Reis MF, Miguel JP, Sampaio C, Aguiar P, Melim JM, Päpke O. Determinants of dioxins and furans in blood of non-occupationally exposed populations living near Portuguese solid waste incinerators. Chemosphere 2007; 67(9): S224-30.
[http://dx.doi.org/10.1016/j.chemosphere.2006.05.102] [PMID: 17240423]

[62]   Reis MF, Sampaio C, Aguiar P, Maurício Melim J, Pereira Miguel J, Päpke O. Biomonitoring of PCDD/Fs in populations living near portuguese solid waste incinerators: levels in human milk. Chemosphere 2007; 67(9): S231-7.
[http://dx.doi.org/10.1016/j.chemosphere.2006.05.103] [PMID: 17215018]

[63]   Reis MF, Sampaio C, Brantes A, *et al.* Human exposure to heavy metals in the vicinity of Portuguese solid waste incinerators--Part 1: biomonitoring of Pb, Cd and Hg in blood of the general population. Int J Hyg Environ Health 2007; 210(3-4): 439-46.
[http://dx.doi.org/10.1016/j.ijheh.2007.01.023] [PMID: 17324622]

[64]   Reis MF, Sampaio C, Brantes A, *et al.* Human exposure to heavy metals in the vicinity of Portuguese solid waste incinerators--Part 3: biomonitoring of Pb in blood of children under the age of 6 years. Int J Hyg Environ Health 2007; 210(3-4): 455-9.
[http://dx.doi.org/10.1016/j.ijheh.2007.01.021] [PMID: 17336151]

[65]   Reis MF, Sampaio C, Brantes A, *et al.* Human exposure to heavy metals in the vicinity of Portuguese solid waste incinerators--Part 2: biomonitoring of lead in maternal and umbilical cord blood. Int J Hyg Environ Health 2007; 210(3-4): 447-54.
[http://dx.doi.org/10.1016/j.ijheh.2007.01.020] [PMID: 17347042]

[66]   Reis MF, Sampaio C, Melim JM, Miguel JP. First results from dioxins and dioxin-like compounds in the population from Madeira Island, Portugal. Part 2 - biomonitoring in breast milk of women living near to a solid waste incinerator. Organohalogen Compd 2004; 66: 2709-15.

[67]   Forastiere F, Badaloni C, de Hoogh K, *et al.* Health impact assessment of waste management facilities in three European countries. Environ Health 2011; 10: 53.
[http://dx.doi.org/10.1186/1476-069X-10-53] [PMID: 21635784]

[68]   The story of the Sint-Niklaas (Belgium) waste incinerator . 1977-2009 .by Fred De Baere, Flemish platform health and environment http://www.gainscotland.org.uk/feature_Sint-Niklaas.shtml

[69]   Ten Tusscher GW1, Stam GA, Koppe JG. Open chemical combustions resulting in a local increased incidence of orofacial clefts. Chemosphere 2000; 40(9- 11): 1263-70.

[70]   Dolk H, Vrijheid M, Armstrong B, *et al.* Risk of congenital anomalies near hazardous waste landfill sites in Europe: the EUROHAZCON study. Lancet 1998; 8;352(9126): 423-7.
[http://dx.doi.org/10.1016/S0140-6736(98)01352-X]

[71]   Dolk H. What is the "primary" prevention of congenital anomalies? Lancet 2009; 374(9687): 378.
[http://dx.doi.org/10.1016/S0140-6736(09)61411-2] [PMID: 19647606]

[72]   Budnick LD, Logue JN, Sokal DC, Fox JM, Falk H. Cancer and birth defects near the Drake

Superfund site, Pennsylvania. Arch Environ Health 1984; 39(6): 409-13.
[http://dx.doi.org/10.1080/00039896.1984.10545873] [PMID: 6524960]

[73]   Dayal H, Gupta S, Trieff N, Maierson D, Reich D. Symptom clusters in a community with chronic exposure to chemicals in two superfund sites. Arch Environ Health 1995; 50(2): 108-11.
[http://dx.doi.org/10.1080/00039896.1995.9940887] [PMID: 7786046]

[74]   Ozonoff D, Aschengrau A, Coogan P. Cancer in the vicinity of a Department of Defense superfund site in Massachusetts. Toxicol Ind Health 1994; 10(3): 119-41.
[http://dx.doi.org/10.1177/074823379401000302] [PMID: 7855862]

[75]   Allan SE, Sower GJ, Anderson KA. Estimating risk at a Superfund site using passive sampling devices as biological surrogates in human health risk models. Chemosphere 2011; 85(6): 920-7.
[http://dx.doi.org/10.1016/j.chemosphere.2011.06.051] [PMID: 21741671]

[76]   Brewer CA. Basic mapping principles for visualizing cancer data using Geographic Information Systems (GIS). Am J Prev Med 2006; 30(2) (Suppl.): S25-36.
[http://dx.doi.org/10.1016/j.amepre.2005.09.007] [PMID: 16458787]

[77]   Copeland G. The role of public health and how boundary analysis can provide a tool for public health investigations: the public health perspective. Spat Spatio-Temporal Epidemiol 2010; 1(4): 201-5.
[http://dx.doi.org/10.1016/j.sste.2010.09.002] [PMID: 22749498]

[78]   English D. Geographical Epidemiology and Ecological Studies. In: Elliott J, Cuzick D, English D , Stern R, Eds. Geographical and Environmental Epidemiology: Methods for Small-area Studies. Oxford: Oxford University Press 1996; p. 3-21.
[http://dx.doi.org/10.1093/acprof:oso/9780192622358.003.0001]

[79]   Gaffney SH, Curriero FC, Strickland PT, Glass GE, Helzlsouer KJ, Breysse PN. Influence of geographic location in modeling blood pesticide levels in a community surrounding a U.S. Environmental protection agency superfund site. Environ Health Perspect 2005; 113(12): 1712-6.
[http://dx.doi.org/10.1289/ehp.8154] [PMID: 16330352]

[80]   Choi AL, Levy JI, Dockery DW, *et al.* Does living near a Superfund site contribute to higher polychlorinated biphenyl (PCB) exposure? Environ Health Perspect 2006; 114(7): 1092-8.
[http://dx.doi.org/10.1289/ehp.8827] [PMID: 16835064]

[81]   Kearney G. A procedure for detecting childhood cancer clusters near hazardous waste sites in Florida. J Environ Health 2008; 70(9): 29-34.
[PMID: 18517151]

[82]   Thompson JA, Bissett WT, Sweeney AM. Evaluating geostatistical modeling of exceedance probability as the first step in disease cluster investigations: very low birth weights near toxic Texas sites. Environ Health 2014; 13(1): 47.
[http://dx.doi.org/10.1186/1476-069X-13-47] [PMID: 24906417]

[83]   Boberg E, Lessner L, Carpenter DO. The role of residence near hazardous waste sites containing benzene in the development of hematologic cancers in upstate New York. Int J Occup Med Environ Health 2011; 24(4): 327-38.
[http://dx.doi.org/10.2478/s13382-011-0037-8] [PMID: 22002323]

[84]   Fortunato L, Abellan JJ, Beale L, LeFevre S, Richardson S. Spatio-temporal patterns of bladder cancer incidence in Utah (1973-2004) and their association with the presence of toxic release inventory sites Int J Health Geogr 2011; 28;10:16

[85]   Alert H. Disease Clusters Spotlight the Need to Protect People from Toxic Chemicals 2011.https://www.nrdc.org/resources/disease-clusters-spotlight-need-protect-people-toxic-chemicals

[86]   Johnston J, MacDonald Gibson J. Indoor air contamination from hazardous waste sites: Improving the evidence base for decision-making. Int J Environ Res Public Health 2015; 12(12): 15040-57.
[http://dx.doi.org/10.3390/ijerph121214960] [PMID: 26633433]

[87]   Navarro J, Gremillet D, Alfan I, Ramirez F, Bouten W, Forero MG. Feathered detectives: Real-time

GPS tracking of scavenging gulls pinpoints illegal waste dumping. PLOSONE 2016; 22;11(7): e0159974.

[88] Sanneh ES, Hu AH, Chang YM, Sanyang E. Introduction of a recycling system for sustainable municipal solid waste management: A case study on the greater Banjul area of the Gambia. Environ Dev Sustain 2011; 13: 1065-80.
[http://dx.doi.org/10.1007/s10668-011-9305-9]

[89] Oke IA. Management of immunization solid wastes in Kano State, Nigeria. Waste Manag 2008; 28(12): 2512-21.
[http://dx.doi.org/10.1016/j.wasman.2007.11.008] [PMID: 18191394]

[90] Ketlogetswe C, Oladiran MT, Foster J. Improved combustion processes in medical wastes incinerators for rural applications. Afr J Sci Technol 2004; 5(1): 67-72.

[91] Botswana Environmental and Climate Change Analysis 29 May, 2008 This Environmental and Climate Change Analysis was written at the request of Sida INEC, Stockholm (att: Rolf Folkesson) by Gunilla Ölund Wingqvist and Emelie Dahlberg at Sida Helpdesk for Environmental Economics, University of Gothenburg as part of Sida-EEU"s institutional collaboration on environmental economics and strategic environmental assessment (https://sidaenvironmenthelpdesk.se/digitalAssets/1683/1683296_environmental-and-climate-change-policy-brief-botswana-2008.pdf)

[92] Kuwoh I, Mochungong P. Environmental exposure and public health impacts of poor clinical waste treatment and disposal in Cameroon. Thesis 2011; PhD series A002: Unit for Health Promotion Research, 2011. (https://www.sdu.dk/-/media/files/om_sdu/institutter/ist/sundhedsfremme/phd+thesis/peter+mochungong+phd+thesis+final+text.pdf)

[93] Bundela PS, Gautam SP, Pandey AK, Awasthi MK, Sarsaiya S. Municipal solid waste management in Indian cities. Int J Environ Sci 2010; 1: 591-605.

[94] Gupta N, Yadav KK, Kumar V. A review on current status of municipal solid waste management in India. J Environ Sci (China) 2015; 37: 206-17.
[http://dx.doi.org/10.1016/j.jes.2015.01.034] [PMID: 26574106]

[95] Seng B, Kaneko H, Hirayama K, Katayama-Hirayama K. Municipal solid waste management in Phnom Penh, capital city of Cambodia. Waste Manag Res 2011; 29(5): 491-500.
[http://dx.doi.org/10.1177/0734242X10380994] [PMID: 20813763]

[96] Chiemchaisri C, Juanga JP, Visvanathan C. Municipal solid waste management in Thailand and disposal emission inventory. Environ Monit Assess 2007; 135(1-3): 13-20.
[http://dx.doi.org/10.1007/s10661-007-9707-1] [PMID: 17492361]

[97] Soares JC, Yunus HS, Kusuma D. Public perception of urban solid waste management in sub district domaleixo Dili-Timor Leste. Geogr Ind Mag 2016; 25(2): 162-80.

[98] Gutberlet J. Gutberlet J. Waste in the city: Challenges and opportunities for urban agglomerations 2017. DOI10.5772/interchopen.72047 (http://dx.doi.org/10.5772/intechopen.72047).

[99] Bandara NJGJ. Municipal Solid Waste Management The Srilankan Case in Developments in Forestry and Environment Management in Sri Lanka. Department of Forestry and Environmental Sciences: University of Sri Jayewardenepura 2008; pp. 93-5.

[100] Ziraba AK, Haregu TN, Mberu B. A review and framework for understanding the potential impact of poor solid waste management on health in developing countries. Arch Public Health 2016; 74(1): 55.
[http://dx.doi.org/10.1186/s13690-016-0166-4] [PMID: 28031815]

[101] Al-Khatib IA, Arafat HA, Basheer T, *et al.* Trends and problems of solid waste management in developing countries: a case study in seven Palestinian districts. Waste Manag 2007; 27(12): 1910-9.
[http://dx.doi.org/10.1016/j.wasman.2006.11.006] [PMID: 17224264]

[102] Shekdar AV. Sustainable solid waste management: an integrated approach for Asian countries. Waste

Manag 2009; 29(4): 1438-48.
[http://dx.doi.org/10.1016/j.wasman.2008.08.025]

[103]   Jambeck JR, Geyer R, Wilcox C, *et al.* Marine pollution. Plastic waste inputs from land into the ocean. Science 2015; 347(6223): 768-71.
[http://dx.doi.org/10.1126/science.1260352] [PMID: 25678662]

[104]   dos Muchangos LS, Tokai A, Hanashima A. Analyzing the structure of barriers to municipal solid waste management policy planning in Maputo city, Mozambique. Environ Dev 2015; 16: 76-89.
[http://dx.doi.org/10.1016/j.envdev.2015.07.002]

[105]   Parameswari K, Padmini TK, Mudgal BV. Assessment of soil contamination around municipal solid waste dumpsite. Ind J Sci Technol 2015; 8: 36.
[http://dx.doi.org/10.17485/ijst/2015/v8i36/87437]

[106]   Tholkappian M, Ravisankar R, Chandrasekaran A, *et al.* Assessing heavy metal toxicity in sediments of Chennai Coast of Tamil Nadu using Energy Dispersive X-Ray Fluorescence Spectroscopy (EDXRF) with statistical approach. Toxicol Rep 2017; 5: 173-82.
[http://dx.doi.org/10.1016/j.toxrep.2017.12.020] [PMID: 29387565]

[107]   Prechthai T, Parkpian P, Visvanathan C. Assessment of heavy metal contamination and its mobilization from municipal solid waste open dumping site. J Hazard Mater 2008; 156(1-3): 86-94.
[http://dx.doi.org/10.1016/j.jhazmat.2007.11.119] [PMID: 18207321]

[108]   Chatham-Stephens K, Caravanos J, Ericson B, *et al.* Burden of disease from toxic waste sites in India, Indonesia, and the Philippines in 2010. Environ Health Perspect 2013; 121(7): 791-6.
[http://dx.doi.org/10.1289/ehp.1206127] [PMID: 23649493]

[109]   Caravanos J, Carrelli J, Dowling R, Pavilonis B, Ericson B, Fuller R. Burden of disease resulting from lead exposure at toxic waste sites in Argentina, Mexico and Uruguay. Environ Health 2016; 15(1): 72.
[http://dx.doi.org/10.1186/s12940-016-0151-y] [PMID: 27339191]

[110]   Habitat UN. Solid waste management in the world's cities.. Water and Sanitation in the World's Cities 2010.https://unhabitat.   org/books/solid-waste-management-in-the-worlds-cities-water-andsa-itation-in-the-worlds-cities-2010–2/

[111]   Liu J, Raven P. China's environmental challenges and implications for the world. Crit Rev Environ Sci Technol 2010; 40: 823-51.
[http://dx.doi.org/10.1080/10643389.2010.502645]

[112]   Alfaia RGSM, Costa AM, Campos JC. Municipal solid waste in Brazil: A review. Waste Manag Res 2017; 35(12): 1195-209.
[http://dx.doi.org/10.1177/0734242X17735375] [PMID: 29090660]

[113]   Agustín B. Hallaron Más Desechos Tóxicosen El Copey. El Tiempo, August 24, 2009. (https://www.eltiempo.com/archivo/documento/CMS-5927747).

[114]   Amador AA. Hazardous waste management solution for Bogota Thesis. The Faculty of the Department   of   Anthropology   San   José   State   University 2010(https://www.sjsu.edu/anthropology/docs/projectfolder/Arjona-Andrea-project).

[115]   Liss G. Zero waste communities.. Retrieved from Zero Waste International Alliance: Working towards a world without waste 2013.http://zwia.org/news/zerowaste-communities

CHAPTER 3

Hazardous Waste And its Associated Health Risks

Rishi Rana and **Rajiv Ganguly**[*]

Department of Civil Engineering, Jaypee University of Information Technology, Waknaghat, District Solan, Himachal Pradesh 173234, India

Abstract: One of the major challenges faced as the result of a growing population and increased industrialization is the management of hazardous wastes that are been generated continuously. Today, hazardous waste presents one of the biggest environmental and public health challenges for any government and society as its safe handling needs a good amount of capital and scientific knowledge. Improper disposal of wastes has led to numerous environmental concerns like air, water and soil pollution. Many research studies have shown that hazardous substances present in the waste produce toxins that affect humans as well as the whole environment. Hence, through analysis of secondary data, this paper tries to throw light on some dimensions associated with it, namely identification and classification of hazardous waste, its harmful impact on health and environment and its disposal and management.

Keywords: Disposal, Environmental and Health Risks, Hazardous Waste, Management.

INTRODUCTION

The waste management issues are a major concern at local, national and international levels. In present times, the economic expansion at an accelerated rate increased the demand of naturally available resources [1, 2]. Pollution from the industries, erosion of soil coupled with deforestation and degradation of lands is causing environmental complications. In the initial eras, the discarding of humans and other forms of waste did not pose a major problem, as the population was very small, and available land for the disposal of waste was huge. Increasing population along with urbanization and industrialization has resulted in the generation of large quantities of waste [2 - 5].

These wastes can be classified as solid waste, liquid waste and atmospheric emissions. Atmospheric emissions and liquid waste owing to their destructive

[*] **Corresponding author Rajiv Ganguly:** Department of Civil Engineering, Jaypee University of Information Technology, Waknaghat, District Solan, Himachal Pradesh 173234, India; Tel: +91-1792-239398; Fax: + 91-179--245371; E-mails: rajiv.ganguly@juit.ac.in, rajiv.phd@gmail.com

Gabriella Marfe & Carla Di Stefano (Eds.)

nature have got much attention. However, solid wastes and efficient management are at the developmental stage.

Solid waste is generally categorized as domestic, industrial, agricultural, constructional, biomedical, commercial and hazardous waste (HW). The management techniques used for handling hazardous waste are basically new and about to introduced in Asian countries, including India. The absence of sophisticated techniques, trained manpower and finances needed for managing hazardous wastes has led to the open dumping of hazardous wastes in India, which creates severe hazards to human beings, mammals and plant life [6]. Though many institutional and regulatory frameworks have been set up by the government [7] but their implementation failed at many levels due to the complex nature of hazardous waste. Fig. (1) shows the intricacy nature of hazardous waste. Hence, this paper tends to understand what constitutes hazardous waste. What are its impacts on health and the environment? What measures have been taken to manage hazardous waste?

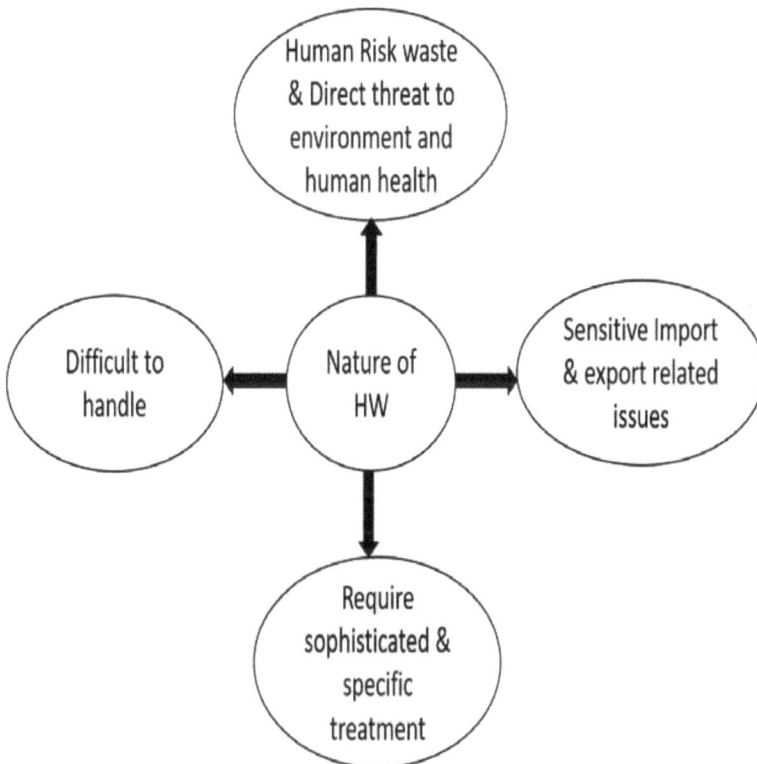

Fig. (1). Complex nature of hazardous waste.

HAZARDOUS WASTE: IDENTIFICATION AND CLASSIFICATION

Hazardous waste has been defined as the waste that poses threat to public health or the environment due to their obstinate behaviour in nature or owing to their properties of biomagnification and toxicity potential [8 - 11]. Due to increased industrialization, a huge amount of hazardous waste arises from different types of industries situated all over in the country. The sources of hazardous waste include the one from the processes taking place in industries, mining extraction, residues of pesticides used in agriculture, *etc.* The various industrial processes and operations produce a huge amount of hazardous waste [12]. The idea about the presence of a huge amount of industrial hazardous waste across the country can be noticed from the data, as shown in Table **1**.

Table 1. State-wise generation of Hazardous waste (CPCB, 2018).

S. No	State/UT	Quantity of Hazardous Waste (MTA)
1.	Andaman & Nicobar	Not Applicable
2.	Andhra Pradesh	282266.4
3.	Arunachal Pradesh	Information not available
4.	Assam	29434.64
5.	Bihar	7629
6.	Chandigarh	2846.892
7.	Chhattisgarh	65186.14
8.	Daman & Diu	Information not available
9.	Delhi	4197.36
10.	Goa	24796
11.	Gujarat	2811925.3
12.	Haryana	58829.43
13.	Himachal Pradesh	29029.38
14.	Jammu & Kashmir	1043.21
15.	Jharkhand	578788.6
16.	Karnataka	336791.6
17.	Kerala	38466.20
18.	Lakshadweep	0.00
19.	Madhya Pradesh	125880.7
20.	Maharashtra	381686.2
21.	Manipur	Information not available
22.	Meghalaya	75.8

(Table 1) cont.....

S. No	State/UT	Quantity of Hazardous Waste (MTA)
23.	Mizoram	0.00
24.	Nagaland	10
25.	Odisha	595697.8
26.	Puducherry	Information not available
27.	Punjab	115490.1
28.	Rajasthan	724663.2
29.	Sikkim	785.472
30.	Tamil Nadu	383189.2
31.	Telangana	277078.5
32.	Tripura	270.19
33.	Uttarakhand	24264.09
34.	Uttar Pradesh	186591.5
35.	West Bengal	85848.74
Total 71,72,762		

A few sources of hazardous wastes can be explosions in incinerators and fire incidents at landfill sites [1, 12, 13]. Others may include biomedical, pathogenic and pathological wastes from various hospitals and radioactive wastes that require special handling constantly. The hazardous waste generated from the industries is a major cause of environmental pollution, resulting in undesirable public health effects [14 - 16].

The classification of hazardous waste is based on factors like nature, its constituents and the composition of the waste. Many times, there is confusion in terms of hazardous substance and waste. A hazardous substance refers to the material having commercial value, while hazardous waste is the material that has been used or no longer in use.

The various acts like Comprehensive Environmental Response, Compensation and Liability Act, 1980 (CERCLA), Toxic Substance Control Act, 1976 (TSCA), Hazardous and Solid Waste Amendments of 1984 (HSWA) and Resource Conservation and Recovery Act, 1976 (RCRA), have given the classification list based on which hazardous waste is classified [17 - 19]. The wastes are classified as:

1) F-list (hazardous waste from non-specific industrial processes);

2) K-list (waste generated from specific industrial processes *viz.*, pesticide, petroleum and explosive industries) and

3) P and U list (commercial sectors, pesticides which have a higher shelf life).

Apart from this classification, the major role is played by the characteristic property of hazardous waste [9, 12, 20 - 23]. On the basis of characteristics, hazardous wastes are categorized as ignitable, corrosive, reactive and toxic [24, 25]. Many times due to the lack of data pertaining to the rate of generation of hazardous waste poses difficulty in the identification and classification of waste. Table **2** describes some common sources of HW.

Table 2. Category and Sources of Hazardous Waste.

S.No	Category of Waste	Source
1.	Radioactive Waste	Hospitals, medical facilities, laboratories of school, colleges, testing centres and universities, thermal and nuclear power plants.
2.	Volatile and Explosive Waste	Construction, ammunition companies, *etc.*
3.	Biological Waste	Biomedical research facilities, medical institutes, clinics and hospitals.
4.	Combustible Waste	Petroleum refineries, dry cleaning and cleaning stations.
5.	Toxic Waste	Waste from pesticides, insecticides, chemical factories, painting shops, nuclear power plants, *etc.*

A combination of any amount of hazardous waste and a solid non-hazardous waste is also considered hazardous in nature [10, 16, 24, 26 - 28]. In recent years, the hazardous wastes from different sources–specific or non-specific- have become more complex and it will tend to become even more complex with the usage of new technologies.

HAZARDOUS WASTE: IMPACT ON HEALTH

The evidences of adverse effects on health due to a large amount and mismanagement of waste, specifically hazardous waste, have been found in both developed and developing countries. Many researchers [8, 10, 11, 14, 17, 19, 22, 29], from their studies, have concluded the harmful impact of environmental exposures on the health of the overall population. The studies have also found that inadequate management of hazardous waste leads to contamination of soil and groundwater [17]. In India, many cases on the presence of lead in water have been reported in areas [3 - 5, 8, 14, 20, 23, 28]. Hazardous waste contains chemicals like polycyclic aromatic hydrocarbons, heavy metals, chlorinated solvents, vinyl chloride, benzene and many chemicals which are being released from the open dumps. The reports like [21, 24, 27], provide a sign of an alliance between open dumping of waste, and people living nearby these particular sites have more undesirable health risks.

All the components present in hazardous waste contain chemicals which are toxic in nature. This can be either chronically or acutely toxic to human health. The acute exposure denotes the exposure to an extremely high concentration of hazardous compounds for short durations, while chronic exposure means exposure to a low concentration of hazardous components for a long duration. The contaminants present in hazardous waste can be chronically toxic to humans as well as mammals owing to the property of bioaccumulation which causes irreversible effects [8, 10, 11, 14, 17, 19, 22, 29]. Many epidemiological surveys have shown increased morbidity and mortality rate among human beings due to hazardous waste. Hazardous waste holds the property of ignitability, which poses a great risk of danger in their everyday handling.

Recent reports show that more than 60% deaths in developing countries are due to an increase in the generation of hazardous waste. Due to a lack of data, these studies aimed to identify and quantify exposure to environmental and health risks. Adverse health impacts due to ill hazardous waste management practices, use of inappropriate waste management techniques and lack of legislation policies represent a potential public health issue in both developed and developing countries. Studies like [9, 12, 20 - 23, 29] revealed that many illegal and unregulated sites of hazardous waste including urban and industrial waste have been in operation that directly causes adverse human health impacts.

The constituents in the hazardous waste tend to exert toxic effects on humans by gaining access in the cells and tissues of the organism. The major routes through which the chemical constituents present in the hazardous waste can enter the human body are inhalation, ingestion and absorption. Therefore, the management of hazardous waste, as well as the related health and environmental risks, have become a crucial global issue [22-26]. Even the health care workers are not aware of the risks which are linked with hazardous waste. Exposure to hazardous waste results in injuries and diseases that can be infectious, causing genotoxicity and are radioactive in nature. All those who come in contact are at potential risk. Many times, workers who are in direct contact with the hazardous waste or even the general public are prone to virus infections like HIV/AIDS, hepatitis B and C.

The health of people, particularly those that belong to the low-income group or the one who is directly linked with the waste, is not only affected by the accumulation of uncollected hazardous waste but also can be conceded by the waste management facilities like open dumps, sanitary landfills or incineration processes. Without the provision of effective equipment and guidance on the handling of these potentially dangerous constituents, hazardous waste tends to pose serious health risks to everyone who comes in contact either directly or indirectly [8, 9, 11, 12, 14, 19, 20 - 23]. There are many uncertainties related to the assessment of health impacts. Therefore, there is a need for independent systematic actions on the current technical information in order to lay down laws by policymakers based on reliable scientific knowledge.

A recent study conducted (CPCB, 2017; The Tribune dated 21-09-2019) stated that five sites in Himachal Pradesh, India have been highly contaminated by the presence of carcinogenic chemicals like cadmium, asbestos, toluene, pesticides and volatile organic compounds like dichloromethane, vinyl chloride and chloroform. The reason behind these contaminations is the dumping of hazardous waste illegally and unscientifically. Therefore, it becomes foremost important for policymakers, regulators, hazardous waste generators as well as other stakeholders to play a critical role in ensuring that these hazardous waste materials are prevented, minimized, collected properly and efficiently treated and disposed.

HAZARDOUS WASTE: MANAGEMENT AND DISPOSAL

The management of a huge amount of hazardous waste is a great challenge for both the municipalities as well for industries. Legal and policy framework has been laid down both nationally and internationally for curbing the crisis. Along with this, there are several worldwide programmes to support the laws [9, 12, 22, 23]. However, an integrated approach for achieving sustainable hazardous waste management (HW), primarily solids, varnishes, paints, solvents and other various manufacturing wastes have not been not enclosed under the Water (Prevention and Control of Pollution) Act, 1974 and the Air (Prevention and Control of Pollution) Act, 1981 in India. To enable authorities to control, handle, transport, treat and dispose hazardous waste (HW) in an environmentally friendly manner is a matter of concern.

The disposal and treatment of hazardous waste have become a significant issue in India, specifically in urban areas. With increasing urbanization, the problems associated with the management of hazardous waste are also rising. In spite of substantial socio-economic development, the guidelines for the hazardous waste management system and its application still remain inefficient.

An effective and integrated approach for sustainable management of hazardous waste includes the following steps: collection, transportation, handling, storage and proper disposal. Lack of understanding, control, guidance and training on management of hazardous waste, the available specific policies also face drawbacks. In countries like India, the techniques for treatment of hazardous waste are unregulated and uncontrolled, and in some cases many times the waste is exported to the developed countries [6, 14, 17, 22, 29]. A major portion of hazardous waste is non-segregated along with non-hazardous waste [5, 7, 13, 19,

20, 25, 28]. Unsystematic and indiscriminate disposal of hazardous waste leads to many environmental problems as well as harm the public health.

In this concern, in 2016, the Ministry of Environment and Forest (MoEF, 2016), enacted new law for disposal and treatment of hazardous waste [5, 7-9, 20, 22]. The law stated that it will be the responsibility of the state governments for the management of hazardous waste. This can be counted as a big step towards the management of HW. The main aim of government policies is to provide a thorough understanding of the problems related to hazardous waste management along with the issues being faced by our society [13, 15].

Therefore, it becomes important to continuously monitor the associated risks with public health and the environment. Along with these, appropriate techniques for minimizing waste by means of reduction at source, reusing, and recycling have to be effectively implemented to decrease the amount of hazardous waste generated and disposed. Fig. (2) depicts a few steps which could be taken to act as protective measures against the ill effects of hazardous waste, which could, in turn, help in reducing the health hazards.

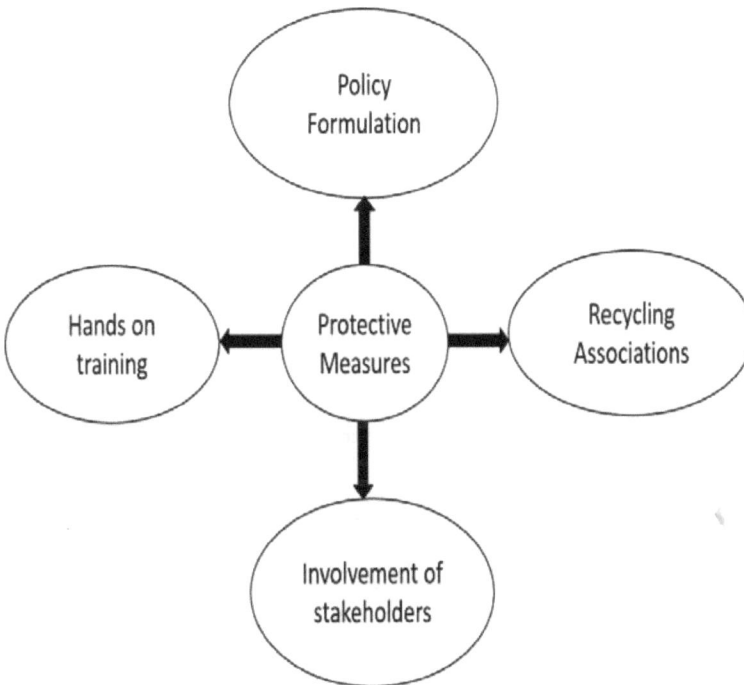

Fig. (2). Protective Measure against hazardous waste.

CONCLUSION

To sum up, from the above analysis, the current situation of hazardous waste, its management and the evaluation of the possible health threats due to the presence of large quantities of hazardous waste is a matter of serious concern. Contaminants released from hazardous waste play major roles in the occurrence of diseases and many evidences have also proven an association between the two factors. There are many mechanisms in the form of laws, regulations and rules to deal with this problem. However, the involvement of all the stakeholders of society is needed to deal with it. There is a need for taking drastic steps and reformation of the existing rules and regulations, especially in the developing countries.

CONSENT FOR PUBLICATION

Not applicable.

CONFLICT OF INTEREST

The authors confirm that there is no conflict of interest to declare for this publication.

ACKNOWLEDGEMENTS

Declared none.

REFERENCES

[1] Al-Khatib IA, Sato C. Solid health care waste management status at health care centers in the West Bank--Palestinian Territory. Waste Manag 2009; 29(8): 2398-403.
[http://dx.doi.org/10.1016/j.wasman.2009.03.014] [PMID: 19398317]

[2] Allaire M, Wu H, Lall U. National trends in drinking water quality violations. Proc Natl Acad Sci USA 2018; 115(9): 2078-83.
[http://dx.doi.org/10.1073/pnas.1719805115] [PMID: 29440421]

[3] ASCEND (waste and environment). The health and environmental impacts of hazardous waste: a report 2015.

[4] Benedict KM, Reses H, Vigar M, *et al.* Surveillance for waterborne disease outbreaks associated with drinking water - united states, 2013-2014. MMWR Morb Mortal Wkly Rep 2017; 66(44): 1216-21.
[http://dx.doi.org/10.15585/mmwr.mm6644a3] [PMID: 29121003]

[5] Centers for Medicare and Medicaid Services. 2018.Historical National Health Expenditure Data https://www.cms.gov/Research-Statistics-Data-and-Systems/Statistics-Trends-andReports/NationalHealthExpendData/index.html

[6] Chatman-Stephens K, Caravanos J, Ericson B, Sunga-Amparo J, Susilorini B, Sharma P. Burden of disease from toxic waste sites in India, Indonesia, and the Philippines in 2010 Environ Health Perspect 2013; 121: 791-6.
[http://dx.doi.org/10. 1289/ehp.1206127]

[7]   Chaves APL, Silva RBD. Environmental diagnosis of hazardous household waste and the family health strategy as liaison for the implementation of a management program in the South of Brazil. Cad. SaúdeColetiva 2015; 23(2): 109-17.
[http://dx.doi.org/10.1590/1414-462X201500020056]

[8]   Central Pollution Control Board (CPCB). Hazardous waste management. India 2017.

[9]   Duggal   A.   2015.A   Public   Health   Legal   Guide   to   Safe   Drinking   Water
http://www.astho.org/Public-Policy/Public-Health-Law/Public-Health-Authority-OverDrinking-Water-Quality-Overview/

[10]   Elsayed E, Frances O. Improper disposal of household hazardous waste: landfill/ municipal waste water treatment. Intech Open 2019.

[11]   Fahad A, Alan R Collins, Elham Erfanian. Drinking Water Quality Impacts on Health Care Expenditures in the United States Working Papers Working Paper 2019-02. Regional Research Institute, West Virginia University 2019.

[12]   Fazzo L, Minichilli F, Santoro M, *et al.* Hazardous waste and health impact: a systematic review of the scientific literature. Environ Health 2017; 16(1): 107.
[http://dx.doi.org/10.1186/s12940-017-0311-8] [PMID: 29020961]

[13]   Inglezakis VJ, Moustakas K. Household hazardous waste management: a review. J Environ Manage 2015; 150: 310-21.
[http://dx.doi.org/10.1016/j.jenvman.2014.11.021] [PMID: 25528172]

[14]   Gutberlet J, Nazim Uddin SM. Household waste and health risks affecting waste pickers and the environment in low- and middle-income countries Int J Occup Environ Health 2017; 23(4): 299-310.
[http://dx.doi.org/10.1080/10773525.2018.1484996]

[15]   Lu JW, Chang NB, Liao L. Environmental informatics for solid and hazardous waste management: Advances, challenges, and perspectives. Crit Rev Environ Sci Technol 2013; 43(15): 1557-656.
[http://dx.doi.org/10.1080/10643389.2012.671097]

[16]   Li L, Wang S, Lin Y, Liu W, Chi T. A covering model application on Chinese industrial hazardous waste management based on integer program method. Ecol Indic 2015; 51: 237-43.
[http://dx.doi.org/10.1016/j.ecolind.2014.05.001]

[17]   Karthikeyan L, Suresh V, Krishnan V, Tudor T, Varshini V. The management of hazardous solid waste in india: An overview. Environments 2018; 5: 103.
[http://dx.doi.org/10.3390/environments5090103]

[18]   Mmereki D, Li B, Meng L. Hazardous and toxic waste management in Botswana: practices and challenges. Waste Manag Res 2014; 32(12): 1158-68.
[http://dx.doi.org/10.1177/0734242X14556527] [PMID: 25432741]

[19]   Sartaj M, Arabgol R. Assessment of healthcare waste management practices and associated problems in Isfahan Province (Iran). J Mater Cycles Waste Manag 2015; 17(1): 99-106.
[http://dx.doi.org/10.1007/s10163-014-0230-5]

[20]   World Health Organization (WHO). Waste and human health: evidence and needs. WHO Meeting Report:. 2016 November; 5–6Bonn, Germany: Copenhagen.

[21]   Zhao J, Huang L, Lee D-H, Peng Q. Improved approaches to the networkdesign problem in regional hazardous waste management systems. Transp Res Part E Logist Trans Rev 2016; 88: 52-75.
[http://dx.doi.org/10.1016/j.tre.2016.02.002]

[22]   Zeller L. Air Modeling Report –"BMWNC Medical Waste Incinerators".Blue Ridge Environmental Defense League 2010.

[23]   Zhao J, Huang L, Lee DH, Peng Q. Improved approaches to the networkdesign problem in regional hazardous waste management systems. Transport Res Part E: Logist Transport Rev 2016; 88: 52-75.
[http://dx.doi.org/10.1080/10643389.2012.671097]

[http://dx.doi.org/10.1080/10643389.2012.671097]

[24]   Li L, Wang S, Lin Y, Liu W, Chi T. Acovering model application on Chinese industrial hazardous waste management based on integer program method. Ecol Indic 2015; 51: 237-43.
[http://dx.doi.org/10.1016/j.ecolind.2014.05.001]

[25]   Karthikeyan L, Venkatesan MS, Vignesh K, *et al.* The management of hazardous solid waste in India: An Overview. Environments 2018; 5: 103.
[http://dx.doi.org/10.3390/environments5090103]

[26]   Mmereki D, Li B, Meng L. Hazardous and toxic waste management in Botswana: practices and challenges. Waste Manag Res 2014; 32(12): 1158-68.
[http://dx.doi.org/10.1177/0734242X14556527]

[27]   Sartaj M, Arabgol R. Assessment of health care waste management practices and associated problems in Isfahan Province (Iran). J Mater Cycles Waste Manag 2015; 17(1): 99-106.
[http://dx.doi.org/10.1007/s10163-014-0230-5]

[28]   World Health Organization (WHO). Waste and human health: evidence and needs. WHO Meeting Report: 2016 November; 5–6 Bonn, Germany: Copenhagen.

[29]   Zhao J, Huang L, Lee D.H, Peng Q. Improved approaches to the network design problem in regional hazardous waste management systems. Transp Res, Part E Logist Trans Rev 2016; 88: 52-75.
[http://dx.doi.org/10.1016/j.tre.2016.02.002]

Biological Effects of Hazardous Waste: Threshold Limits of Anomalies and Protective Approaches

Amal I. Hassan and **Hosam M. Saleh**[*]

Radioisotope Department, Nuclear Research Center, Atomic Energy Authority, Dokki, 12311, Giza, Egypt

Abstract: Many of the chemicals and prescription drugs utilized in health care foundations are risky substances. Therefore, risk assessment sounds to be one of the tight evolving appurtenances for developing convenient management methods which will assist in hazardous waste management decision.

Some dangerous substances induce venomous effects on humans or the surroundings, once casual exposure, these venomous effects are described as acute toxicity. As for the prolonged exposure to these dangerous substances, it is called chronic toxicity.

Furthermore, one of the world health organization's functions is to supply objective and reliable data and recommendation within the field of human health, a responsibility that it accomplishes to some extent through its publication programs. Hazardous waste reduction provides regulatory benefits as well.

The exposure to noxious substances has a range of effects on biological systems. The present review will highlight the evidence of the association between exposure to hazardous wastes, and threshold limits of the susceptibility of anomalies, besides, the protective proceedings.

Keywords: Biological Effects, Congenital Anomalies, Hazardous Waste, Health, Threshold.

INTRODUCTION

Some harmful substances have toxic biological effects on humans or the environment after a one-time accidental exposure. These toxic effects are referred to as acute toxicity [1].

[*] **Corresponding author Hosam M. Saleh:** Radioisotope Department, Nuclear Research Center, Atomic Energy Authority, Dokki 12311, Giza, Egypt; Tel: +20 1005191018; Fax: +202 37493042; E-mail: hosamsaleh70@yahoo.com

Gabriella Marfe & Carla Di Stefano (Eds.)

Perilous wastes encompass a large variety of completely different environmental pollutants that include many exposure methods, looking for patterns of those materials, hydrogeological factors and meteorology. These limitations tend to reduce, rather than overestimate, the magnitude of health impacts, thereby mitigating evidence of specific patterns of exposure to adventurous waste [2].

The potential health effects of toxic substances usually include four specific categories: potential cancer, genetic toxic mechanisms, developmental effects, and organ/tissue effects, besides, an assessment of the conditions under which these toxicities may appear in exposed humans [3].

All recognized carcinogens had yielded positive results in animal models. Therefore, in the absence of sufficient accessible data, it is reasonable to consider the factors or compounds that can be inferred from carcinogenicity and which pose a potential danger to humans in experimental animals [4]. Thence, chemicals that induce tumors in animals, are supposed to cause tumors in humans also. The most appropriate biological examinations for rodents are those that test exposure pathways most relevant to human exposure pathways, such as inhalation, mouth, skin and other exposure methods. Since biological tests can be combined, so it is recommended to connect these tests with the mechanism of biomarker and genetic studies to understand the mechanism (s) of toxicity and/or carcinogenesis [5].

Evaluation of dose-response is essential for the quantitative relationship between factor exposure and the occurrence of a negative response. The procedures used to determine the relationship between dose and response to carcinogens and non-carcinogens vary. For carcinogens, a non-threshold, zero, dose-response relationship is used when there is a known or assumed risk of a negative response at any dose above zero. Non-threshold toxic substances include toxic substances for genetic diseases, carcinogens of toxic genes, and genetic developmental toxic substances. For non-carcinogenic agents, a non-zero threshold is used to evaluate toxic substances known or assumed to produce no adverse effects below a given dose or dose rate. Threshold toxins include non-gene carcinogens, non-gene developmental toxicities, and organ/tissue toxins [6, 7].

The standard terminology surrounding hazardous materials can be best described as confusing. Several possible terms seem similar; however, they typically have completely various restrictive meanings. When you are unsure about the meaning of a term, remember that legal definitions are found in individual regulations. When dealing with hazardous materials, several key definitions must be addressed and understood.

Hazardous materials are groups of them that have properties capable of causing harmful effects on human health, besides, the safety of the environment (Table **1**).

These toxic wastes include hazardous chemicals and medical waste; they may be acute or chronic when exposed to [6]. The results of previous studies demonstrated that the vulnerability to these wastes may cause either sudden death or illness shortly after exposure (Table **1**).

Table 1. Summary of treatment and disposal options for biomedical waste.

SI. No.	Waste category type	Treatment and disposal option
1	Animal waste	Incineration/deep burial
2	Chemical waste	Treatment by chemical for liquids. Secured landfill for solid
3	Genotoxic	Destruction/Incineration and disposal in secured landfill
4	Incineration ash	Disposal in municipal landfill
5	Microbiology and biotechnology Wastes	Local autoclaving/microwaving/incineration
6	Pathological waste	Incineration/deep burial
7	Pharmaceutical waste	Destruction/Incineration and disposal in secured landfill
8	Pressurized containers	Return to suppliers/Controlled destruction
9	Radioactive waste	Concentrate and contain or dilute and disperse
10	Sharps	Disinfection and mutilation/shredding
11	Soiled waste	Destruction/incineration and disposal in secured landfills
12	Waste with heavy metal	Heavy metal recovery

THE THRESHOLD OF HAZARDOUS MATERIALS IN THE BIOLOGICAL SYSTEMS

Health risks from chemicals and hazardous wastes resulting in have become a critical concern for the public. It is now generally recognized that chemicals in the human environment need to be better controlled [8]. Recent laws have been enacted in many countries guiding the manufacture, distribution, and removal of toxins. In many cases, they also organize experimental toxicity testing of chemicals [9].

The scientists determine levels of exposure to humans that will lead to noticeable symptoms through two types of studies. Epidemiological studies use data on how toxic substances affect humans and also study statistical models of disease in groups of individuals [10]. This modality of study may compare the number of workers exposed to a particular substance with lung cancer to those in the rest of the population [11]. This modality of study may compare the number of workers exposed to a particular substance with lung cancer to those in the rest of the population. The value of epidemiological studies is usually limited, owing to a lack of qualitative information on the enumeration of population or the concentrations of chemicals that may exist simultaneously, and therefore the effects are difficult to translate. These studies on laboratory animals serve as a basis for predicting the toxic consequences of chemicals on humans when information about the effects of those synthetic substances on the health is scarce. For most chemicals, doses below the threshold limit may not exert a toxic effect. Then other clinical studies test the effects of concentrated doses of substances on animal tissues [12]. A basic principle of toxic chemicals is that the toxic effect intensifies with increasing the dose. Theoretically, there is a threshold of exposure for each substance, below the threshold; the dose is so small that no antagonistic effect will appear [13]. With increasing the dose, it has an effect, but the organism can be compensated by internal healing, and a permanent injury will not occur. Moreover, it has been shown through animal studies, that there is an irreparable dose as a direct result of apparent injury and progressive disease [14]. The threshold unit for the negative effect of a chemical is defined as the concentration or dose that exceeds detectable adverse effects and reduces exposure conditions. This threshold is derived from the level of adverse effect observed [14].

Through experiments, scientists try to determine a certain dose of the chemical (in mass per kilogram of body weight) that will lead to the death of half of the test animals, and this is called the lethal dose or LD50. Clinical signs and symptoms are carefully monitored and behaved. In such trials, it is usually possible to determine the maximum allowable and fatal doses. They also attempt to repair the point at the other end of the curve where there is no noticeable effect of the material on the animal, this meaning that the negative impact cannot be observed [15]. Toxicology is based on the principle of a relationship between a dose and response, and researchers estimate that a minimum dose in which no response occurs. For example, a therapeutic patient's dosage is that prescribed by the doctor, so lower doses are unhelpful and do not have any therapeutic effect and do not benefit the patient [16, 17]. The Toxic higher doses are causing organ damage, and lethal doses are usually more dangerous than toxic doses, which are potentially life-threatening. Allergic reactions or side effects are not toxic; they are not dose-related, they are related to the nature of the body and occur regardless of the dose the person has taken, while toxic reactions are caused by the effects of the chemical on the cells.

The value of the lethal dose of the toxin is influenced by many factors such as animal species, age, sex, diet, the degree of pre-poisoning food deprivation, temperature, and test procedures [12]. The studies mitigate the impact of these factors on the study to make them as accurate as possible [13]. As well as, the harmful effects of a toxin vary from one organism to another. The toxicity may be extremely in mice, but less to another animal. The lethal dose varies radically from person to another, depending critically on the person's structure. The effective dosage varies remarkably, according to the possible route of the administration. Consequently, some substances are noxious when administered through the skin, while others are more toxic if taken by mouth or intravenously. Furthermore, the potential toxicity in the time in which the toxic substance was properly administered may have a potent effect on the toxicity level, particularly through the skin or inhalation. It should be noted approvingly that the lethal dose is properly expressed in milligrams per kilogram of body weight of the exposure to the toxic material. The most minimal dose positively affecting animals, will also have an impact on humans. Also, scientists investigate the effect of matter on susceptible populations where statistics are available. The best way to accurately estimate lethal dose values for humans is by extrapolating results from human cell transplantation. One form of lethal dose measurement is the use of animals like mice or rats, turning them into a dose per kilogram of biomass, and extrapolating human standards. The biology of experimental animals differs in important respects from those of humans. It is worth noting the response of rat tissue to the Sydney spider venom is fifty times lower than that of humans. Another uncertainty associated with the LD50 concept is that most LD50 data is obtained from an acute exposure test (single dose) rather than chronic exposure [13]. The induction of these studies is complicated by the fact that chemicals are sometimes distributed differently in the body when the exposure is chronic; for example, a different target member may be attacked, or the material is emptied more easily. Choosing 50% lethal as a standard avoids the possibility of ambiguity in carrying out measurements at maximum limits and reduces the number of tests required. However, it also means that LD50 is not the lethal dose for all substances; lower doses may cause death, while survival with doses higher than LD50. Measurements such as "LD1" and "LD99" (the dose required to kill 1% or 99%, respectively, of the test sample personnel) are sometimes used for specific purposes [14].

The effects of exposure to venomous substances on human health represent 2 denominations: short and semi-permanent effects. Short effects (or acute effects) have a comparative incipience onset (usually minutes to days) once short exposure to relatively high concentrations of the material (acute exposures) (Table **2**).

If the effect of the chemical is limited to the contact area, it is recognized as the local effect, and the substance is absorbed into the blood circulation, it will be transported to various organs of the body and cause a systemic effect [17]. The native effects occur at the positioning of contact between toxins, and therefore the body. These sites include skin or eyes, the lungs by irritants, and the gastrointestinal system through the administration of destructive substances [18]. The systemic consequences are those that occur if the toxic substances are absorbed into the body from the foremost point of contact, they're relocated to other parts of the body, inflicting harmful effects in sensitive organs (Table **2**). Some chemicals cause severe (adverse) effects shortly after exposure, while other chemicals produce chronic influences, such as cancer, that can occur only 10-20 years after exposure [19].

Table 2. Categorization of agents determined of the toxicity.

Type	Examples
Concerning to chemicals	Composition (salt, free base,*etc.*); physical characteristics (particle size, liquid, solid, *etc.*); physical properties (volatility, solubility, *etc.*); presence of impurities; break down products; carrier.
Exposure factors	Dose; concentration; route of exposure (ingestion, skin absorption, injection, inhalation); duration.
With respect to the person exposed	Heredity; immunology; nutrition; hormones; age; sex; health status; preexisting diseases.
Environment	Carrier (air, water, food, soil); additional chemical present (synergism, antagonism); temperature; air pressure.

The lowest dose of the substance is introduced by any route, except inhalation, over a certain time called the low toxic dose (TDLO). These very low doses of toxic compounds in the environment contribute to a wide range of human health problems, including obesity, diabetes, cancer, cardiovascular disease, infertility, and other sexual development disorders [20]. In an experiment on male prenatal mice exposed to very low levels of artificial estrogen, it was observed that they formed heavier prostate glands than non-susceptible mice, making them later susceptible to prostate diseases, including cancer. Remarkably, high doses of synthetic estrogen did not produce the same effect. These studies helped to stop the use of Bisphenol A in breastfeeding bottles and sip cups for children. It also inspired a large crowd of researchers to explore endocrine-related effects in animals exposed to low levels of bisphenol A and other hormones. Studies have focused on this toxic substance by finding that early exposure to Bisphenol A may alter the development of mammary glands in mice and rats and stimulate the growth of estrogen receptors, leading to pre-cancerous tumors and non-metastatic cancers [21]. Another study confirmed the considerable risk of this substance by exposing human pancreatic cells to Bisphenol A, hence mild correlations were verified between dose levels and glucose metabolism, a major risk factor for diabetes and obesity [22]. The absorption of chemicals in the body doesn't need to produce negative effects. Because the body is adequately equipped with multiple mechanisms to defend*versus* harmful substances. The substances that are absorbed by the body are fatty (dissolve in fat) are more difficult to put out of the body. These substances undergo liver detoxification, called biotransformation, which in turn transforms the materials and forms metabolites. These metabolites are similar to the original constituents but are more soluble in water, and therefore easy to release. In general, their toxicity is much lower than the authentic material. Sometimes metabolites are more toxic than the original substances [23]. If the chemical dose has negative effects, the damage may be reversible or irreversible. The reversible effects are characterized by the fact that the change from the natural structure or function induced by the chemical returns to normal limits when exposure ceases. Table **3** displays the effect of low toxicity, and the lethal dose (LD50) is expected to cause the death of 50 percent of an entire defined experimental animal population whether chronic or acute doses [21]. Damage from non-reversible effects remains or increases even if the exposure is prevented. Certain effects of toxic chemicals are irreversible, including neurological diseases, cancer, cirrhosis, and emphysema [23].

Table 3. Conditions for dose response.

Category	Exposure Time	Route of Exposure	Toxic Effects	
			Human	**Animal**
TDLO	Acute or chronic	All except inhalation	Any nonlethal	Reproductive, Tumorigenic
TCLO	Acute or chronic	inhalation	Any nonlethal	Reproductive, Tumorigenic
LDLO	Acute or chronic	All except inhalation	Death	Death
LD50	Acute	All except inhalation	Not applicable	Death (statistically determined)
LC50	Acute	Inhalation	Not applicable	Death (statistically determined)

Some hazardous substances will prompt way further severe consequences like chromosomal abnormalities that lead to genetic mutations, cancer, physiological malfunctions (*e.g.*, reproductive impairment, kidney diseases, *etc.*), physical deformations, and congenital deformities [22, 23]. Hazardous waste embraces a considerable variety of potential contaminants, that affect numerous environmental matrices, and include many possible ways of exposure, properly looking on forms of junks, hydrographical and atmospheric factors. These restrictions tend to underestimate instead of overestimating the ultimate consequences of health, thereby reducing striking proof of forms of potential exposure to unsafe wastes [23]. Therefore, the objective of appraising the safety of chemicals in food, air, and water is to determine the potential daily intake (PDI) to assess the possible daily intake of chemicals that exist for a lifetime without causing a predictable health risk. Because in most cases there is insufficient information from humans to allow calculations of this daily intake, but the results of animal studies should be measured to provide a significant indication in humans.

Although the PDI can be exceeded for a short period, it is impossible to generalize the probable duration of the time interval that causes some potential problems. The likelihood of adverse effects depends on factors that vary appreciably from chemical to another. The biological half-life of the chemicals or the significant time the body needs to get rid of these chemicals, the nature of the toxicity and the amount of exposure exceeding the PDI are all critical [24]. Predominantly, carcinogenic chemicals through interaction with genetic material are considered non-threshold substances. So, there is a distinct possibility of imminent harm or considerable danger at any level of exposure. As follows, the presence of PDI is considered inappropriate, and mathematical models are used to estimate risk at remarkably low vulnerability levels that may undoubtedly exist in ordinary exposure. On the other hand, there are carcinogens capable of invariably causing tumors in animals or humans without interacting positively with the genetic material, but they intensely affect through an indirect mechanism. Several scientists believe that the threshold dose is less than the expected negative effects present in these non-toxic carcinogens [24].

Concerning the underlying mechanism of cancerous growth, each carcinogen must be evaluated at the basis of each case, taking into account the evidence of gene toxicity, varied types of cancer, and the relationship between humans and tumors observed in laboratory animals.

Animal studies to determine the likelihood of cancer in humans are believed to be thoroughly reliable. All known human carcinogens, which have been studied extensively on experimental animals, have been shown to cause tumors in one or several species of animals.. All known human carcinogens, which have been studied extensively on experimental animals, have been shown to cause cancer in one or several species of animals. For many factors that have caused cancer, such as aflatoxin, tobacco, coal tar, and vinyl chloride, this has been demonstrated in experimental animal models before epidemiological studies confirm the ability of these chemicals to stimulate cancer in humans. Although this evidence has not established that all the factors that cause cancer in experimental animals (EA) necessarily cause cancer in humans, it is generally believed that in the absence of sufficient information in this regard, it is prudent to consider the factors or chemicals that induce cancer in EA. As if they cause cancer risk in humans. Based on this essential principle, the International Agency for Research on Cancer (IARC), in its comprehensive assessment of cancer to the following groups as shown in Table **4**:

Table 4. Classification of chemicals by International Agency for Cancer Research (IACR).

Groups	Chemicals
GI: Carcinogen to man	Aflatoxins, arsenic, arsenic, benzene and tobacco smoke
GIIA: They may be carcinogenic to humans	Cadmium and its compounds, formaldehyde, polychlorinated 2-phenol, vinyl bromide and benzo (a) pyrene,
GIIB: The agent is likely to be carcinogenic to humans	Sulfonate, carbon tetrachloride, DDT, hexachlorobenzene, urethane and saccharin
GIII: Uncategorized based on carcinogenicity	Acrylic fibers, Aldrin, aniline, captain, cholesterol, Bis-Aldrin, lamps, leaves, polyvinyl chloride and vinyl acetate
GIV: It may be non-carcinogenic to humans	Caprolactam

Group I: Carcinogen in humans, this classification is used when there is sufficient evidence that the substance is carcinogenic in humans.

Group IIA: The factor may be carcinogenic to humans, this classification is used when the evidence of cancer in the population is limited, and there is reliable confirmation of cancer in experimental animals.

Group IIB: The agent is likely to be carcinogenic to humans. This classification is used when there is only limited evidence of human carcinogenicity and less convincing evidence of carcinogenicity in experimental animals.

Group III: Factor not classified on the basis of carcinogenicity, this classification is frequently used when the evidence of carcinogenic in humans is insufficient and is extended in experimental animals.

Group IV: Factor may not be carcinogenic to humans; this classification is used when the chemical is fully examined and is not believed to be able to cause cancer in both humans and experimental animals [25]. Also, from Table **5** we note an assessment of the risk of carcinogens and estimated based on human studies [25].

Hepatitis B and C viruses, alcohol consumption, tobacco smoking, and aflatoxin are the most common risk factors for liver cancer. The liver is a key organ in the metabolism, and it processes nutrients that have been absorbed into the blood from the gastrointestinal tract or through other pathways. The liver is used to disassemble nutrients means that it is particularly susceptible to any toxins within the body. The decomposition of nutrients occurs in the liver, which means that it is particularly vulnerable to any toxins inside the body. Liver cells can regenerate after damage from toxic substances, the most common being alcohol. However, continuous absorption can bypass the process of cell regeneration, and cause permanent liver damage. Cirrhosis is a precursor to liver tumors, so long-term liver destruction from industrial toxicity, promptly makes workers susceptible to liver tumors [26]. The liver is a protective organ in itself, as the detoxification processes performed by the liver usually convert potential toxins into safe substances and sometimes *vice versa* [27]. Possible responses to materials may vary by different species. Dioxins cause severe liver damage and death in guinea pigs (experimental mice) but cause skin disease (chlorine count) in monkeys and humans. Arsenic causes cancer in humans, but not in experimental animals. Small doses of atropine cause human death, but do not kill rabbits. A possible response to materials may also vary among individuals. Some smokers may progressively develop lung cancer, while others do not. Penicillin does not cause any possible harm to most people, but it leads to allergic reactions in some. The toxic effect of a substance depends on its physical form, proper dosage, and course of entry, absorption, distribution, metabolism, and excretion [28]. When grinding the rigid materials, it will produce dust that may be inhaled and absorbed or may contaminate the skin. Also, liquids may be swallowed, gases, vapors, fog, and organic aerosols may be inhaled [27].

Table 5. Risk of carcinogens based on human studies.

Substance	Risk of unit N	Site of malignancy
Acrylonitrile	2×10^{-5}	Lungs
Arsenic	4×10^{-3}	Lungs
Benzene	4×10^{-6}	Blood
Chromium VI	4×10^{-2}	Lungs
Nickel	4×10^{-4}	Lungs
Phenyl chloride	1×10^{-6}	Liver and other sites

N: Estimation of cancer risk for lifetime exposure at 1 $\mu g/m^3$ concentration.

The three main entry points for potential toxins into the body are inhalation, skin, and ingestion. Ingestion is the most dangerous in environmental toxicology. During the evolution, mechanisms in the gut to properly regulate the nutrient absorption of essential substances arose. Toxic elements may need competing to absorb just a tiny fraction of any doses had swallowed (usually 10% or less) [29]. Particles less than 10 microns in diameter may be able to reach the alveoli, if they are soluble, about 40% is absorbed. Insoluble chemicals are relatively safe, for example, lead sulfide, while lead carbonate is highly soluble and can lead to rapid poisoning [30]. The considerable risk of large inhaled particles decreases, as the efficiency of absorption decreases above the airway. It is significant to remember that the lung is not exclusively responsible for the assimilation of substances into the body. It is also a target organ, and inhalation is responsible for 90% of industrial intoxication. Substances that do not absorb into the body can remain in the lungs and cause damage to them. The skin as another possible way to voluntarily enter the toxic substances in the body, there is no selective absorption [31]. The fat-soluble compounds are already absorbed as with organic solvents. Percutaneous absorption happens for benzene nitrate, phenol, mercury, and aniline, and the absorption of phenol may be fatal. Once substances enter the body, they can spread around the body by supplying blood destined for plasma proteins, or red cells, concentrations in each member may vary [32]. Other toxic substances may dissolve, or turn to fat, only fat-soluble substances can cross the blood-brain barrier. The substances that are distributed throughout the body tend to metabolize. The only site of metabolism is the liver, as mentioned above, but the kidneys, lungs, and skin can also metabolize certain chemicals [31, 32]. The conventional detoxification process involves oxygenation stages followed by conjugation with glucuronic acid. The rate of metabolism depends on the rate of absorption (the ability to absorb water-soluble compounds is lower than that of fat-soluble), and the extent of protein binding (it reduces the concentration at the metabolic sites). Enzymatic systems develop poorly at a very young age and are therefore metabolized more slowly. The liver converts substances from hydrophobic forms into hydrophilic so, they can be excreted easily through the kidneys, or bile [30]. With genotoxic, carcinogenic agents, which are likely to be a principal cause of concern at low exposure levels, the initial adverse effect is thought to be the induction of a break in the double helix DNA of stem cells or their immediate progenitors. It is becoming feasible to culture such cells *in vitro* and this may be a promising approach to assessing the impacts of such agents either singly or in combination, *e.g.* by studying the induction of mutations, chromosomal aberrations, genomic instability or other deformities of DNA misrepair. However, this addresses only the initial induction of effects at the sub-cellular level. Additional modeling, supported by data, is required to interpret these results in terms of likely increases in cancer induction. Tumour initiation, proliferation and progression all need to be addressed.

Multi-stage models of carcinogenesis may be useful in this context [33]. In general, toxicological assessment criteria for non-radiological substances are based on limiting intakes to those that either fall well below a threshold above which deleterious effects can occur, or that corresponds to a tiny risk of non-threshold impacts such as cancer. For carcinogens and mutations, curves don't show any threshold, a minimum of within vary of doses accessible for the experiment. For very low concentrations, it is assumed that the curve (in this case straight-line) continues linearly until the parent, thus below the threshold. According to this hypothesis, even a single molecule can cause cancer (Fig. **1**).

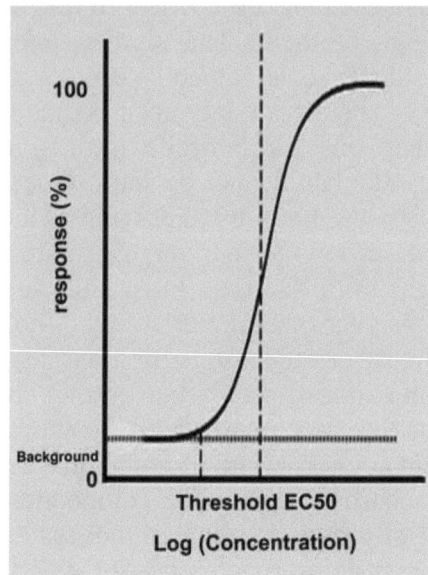

Fig. (1). Curve of dose- response [32].

Below that threshold value, exposure to a toxicant should have no measurable effect. Some feedback to toxicants, such as the occurrence of tumors, occur spontaneously in control groups [34]. In the notable absence of exposure, this rate of occurrence is considered the "background" level. If an incremental dose (ID) increases this background rate of response, then the relationship between the ID and response can be fitted linearly.

CHEMICALS OF PUBLIC HEALTH CONCERN AND PROPERTIES OF HAZARDOUS WASTES

The major inorganic elements including aluminum, arsenic, cadmium, and, chromium, copper, cyanide, iron, lead, besides, mercury, nickel, silicates, tin, zinc, and their compounds are plentiful naturally. The most dangerous inorganic elements include arsenic, cadmium, chromium, copper, lead, mercury, and zinc [35]. For centuries, a considerable number of these toxicant metals are relied upon as raw materials for industrial production (Fig. **2**). A Notable advance in science and technology, and industrialization have contributed mightily to a final product vary tremendously of applications of those dangerous metals and their trace parts. The use of inorganic compounds in industries has been greatly exaggerated over the past two centuries [36]. These substances have harmful effects as they accumulate in living tissues and invariably cause severe toxicity and cell death. Organic halogens, are compounds whose organic molecules typically contain some halogen atoms such as chlorine or fluorine. These materials have multiple uses in all aspects of agricultural, domestic, and industrial life, such as Freon compounds, and some insecticides such as DDT, Lindane, polychlorinated biphenyls, and dioxins. These substances are highly toxic and have slow decomposition that lasts for decades. Experts place these compounds at the top of the list of toxic and hazardous substances found in industrial waste in developed countries. Research on experimental animals has shown that these substances are stored in the body and especially in adipose tissue. The marked increase in these serious materials intensities in the body of the organism leads to possible cancer [36]. There are some other organic compounds containing halogens that have been widely used for many purposes, including these polymer materials known as polyvinyl chloride. Its risk comes from the fact that it contains only a small percentage of free vinyl chloride, which causes cancer. Cyanide compounds are considered to be highly hazardous and highly toxic substances due to their impact on human health and the environment [37]. Cyanide compounds are used in electroplating, metal cleaning, rubber, silver polishing, and rodenticides. Cyanide compounds exist in three forms, solid, gas, and liquid. These compounds can be classified into two types; firstly, simple cyanide compounds, which are chemical compounds in which the cyanide ion binds to a basal fissure (sodium, potassium) or a metal fissure. Sodium and copper cyanide are dissolved in water and are ionized to ions. Cyanide ions are especially dangerous when the pH< 6 because these ions react with hydrogen rapidly forming hydrogen cyanide, which is also a very toxic gas. Secondly, complex cyanide which has different forms where it is associated with a base slit and heavy metals (copper, nickel, cadmium, *etc.*). Complex cyanide compounds are widely used in electroplating. The spread of cyanide compounds in the environment through their release to air, surface, and groundwater water or soil leads to pollution and toxic hazards to humans and other organisms. Where hydrolysis of cyanide by hydrolysis or photolysis leads to the production of free cyanide, which binds with the hydrogen ion and produces Poisonous gas, spread in the environment [37].

Since the Industrial Revolution, the production and consumption of heavy metals worldwide, such as mercury, lead, cadmium, beryllium, iron, and copper has increased dramatically.

Heavy metals are not biodegradable and remain toxic as long as they exist. Most of these metals also bioaccumulate because they work across the food chain or food levels. They are used for many matrices for industrial, commercial, agricultural, and household purposes. Serious diseases have been linked to acute and chronic exposure to these elements and their compounds (Table 6). For example, mercury is a known neurotoxin, and arsenic may cause chronic exposure to hyperpigmentation, carcinoma, and the lead is associated with neurological disorders and mental retardation, especially among children [38].

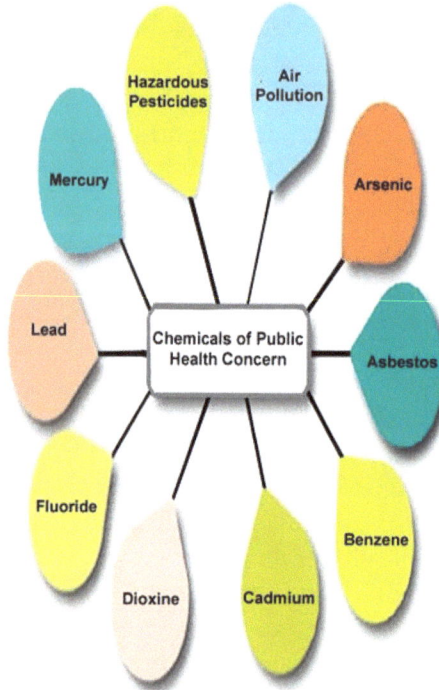

Fig. (2). The chemicals of major public health concern.

Table 6. Some existing natural elements and their effects on human health.

Metal	The influence	Source of exposure
Cadmium (Cd)	Kidney Poisoning	Occupational exposure resulting from inhalation of cadmium fumes and food contamination.
Chromium (Cr)	Skin infections	Occupational exposure and wearing jewelry containing chromium
Lead (Pb)	inhibits the production of hemoglobin in the blood: causes kidney loss of function, mental impairment, (especially more sensitive children)	Occupational exposure, children with direct contact with waste and filth, inhaling lead-containing motor fuels and eating foods containing lead-based dyes

NUCLEAR OR RADIOACTIVE WASTE

Nuclear or radioactive waste refers to the materials that contain radionuclides, but are considered to be no longer useful. According to the International Atomic Energy Agency (IAEA), radioactive waste is any material for which no use is foreseen, which contains radionuclides at concentrations higher than the values deemed acceptable by the competent authority in materials suitable for use not subject to control. Radioactive waste can be found in solid, gaseous, liquid, or sludge states, and its radioactivity levels may vary [39]. However, the safe disposal of radioactive waste is one of the major problems that serve a notable role in asserting nuclear technology applications [40].

RELATIONSHIP OF HAZARDOUS WASTES AND CANCER MORTALITY

Known carcinogens are widely distributed in the environment, and they are found in air, water, food, and soil [41]. Risk assessment is very difficult that typically needs tremendous information. For example, assessing cancer-specific risk requires consideration of factors like dose, duration, possible exposure, and enhancement or inhibition of the impacts of other agents [41]. The risk assessment also involves knowledge of dose-response relationships. Besides, for environmental exposures (non-occupational), assumptions about the magnitude of impacts at low doses cause significant problems in risk assessment. Several studies have shown a varying increase in lung cancer deaths for men and women living in urban centers compared to those living in rural areas, even after controlling the effects of cigarette smoking. Efforts have been made to determine the risk of lung cancer from air pollution using Benzo[a]pyrene ($C_{20}H_{12}$) as an indicator of air pollution [42].

Furthermore, the presences of organic chlorinated compounds have an in-depth relationship with cancer, like vinyl chloride [43], and beta-hexachlorocyclohexane, β-HCH [43], and of serious metals, as well as arsenic [44]. Different studies [45, 46] regarded areas during which biomonitoring detected the presence of dioxins in the breast milk of mothers [47]. Within the latter studies, the synergistic effects between, utterly different risk factors, additionally as disease ensuing from B virus and dangerous waste contaminants is additionally fairly hypothesized [48]. Another review found bladder cancer among those rumored in additional than one study [49]. Variety of things are known by IACR as connected to bladder cancer with sufficient proof, as well as tobacco smoke, arsenic and its compounds, and activity exposure that happens in industries involving the assembly and/or use of aluminum, ammonia, purple and rubber coatings.

The proof is classed as restricted to activity exposure of hairdressers, printing and textile process, and chemical agents surely chlorofluorocarbons, like 4-chloro-ortho-toluidine and anthelmintic [50]. A particular association between bladder cancer and benzo (a) pyrene was reportable elsewhere [42].

REPRODUCTIVE EFFECTS

Reproductive toxicity implicates counteractive effects on sexual performance and male and feminine fertility, likewise as any effect that interferes with normal development before and after childbirth (also called evolutionary toxicity). The physiology of the reproductive system in men differs from women, but in both cases the reproductive system is controlled by hormones that control the development of the genitals and the formation of sperm in males. In females, hormones control the development of the reproductive organs, the female reproductive cycle, and the preparation of the uterus for pregnancy and diuresis [51]. Hormones also play an essential role in pregnancy and fetal development. Under normal conditions, it is accurately estimated that one in five couples who infertile, more than a third of fetuses die in early stages, and approximately 15% of pregnant women are spontaneously aborted. Almost 3% of newborns have birth defects. Not surprisingly, synthetic chemicals interfere constructively with several biological processes of the reproductive system in both males and females [51].

Additionally, some site-specific studies have also been conducted. Hassan *et al.* [52] reported that lead (Pb) about ¼ LD50 is a gonadotoxic with a tendency of suppressing semen characteristics and testosterone levels of animals.

Previous studies have shown that lead can pass through the testicular blood barrier, accumulates in the testicle and/or epididymis and have an effect on bacterial cells at completely different levels of differentiation (spermatogonia, primary spermatocytes, spermatids or spermatozoa) [53].

There are three goals of reproductive toxicity where they can directly affect the central nervous system and alter hormone secretion (such as synthetic steroids). The gonads (ovaries and testicles) are also the target of drugs and chemicals, especially cancer chemotherapy drugs. Reproductive toxins also prevent or alter sperm. The results of such toxic effects include infertility, infertility, increased fetal death, increased infant death and increased birth defects. Chemicals that cause an increase in birth defects are called teratogens. Adverse effects may occur at pre-fertilization (either parent), during pregnancy, or from birth until sexual maturity. Evolving adverse effects may be detected at any point in the organ's lifetime. The greatest manifestations of evolutionary toxicity include: (a) death of the developing organ. (B) Abnormal structure. (C) Variable growth. (D) Functional destitution. Exposure to chemicals during pregnancy may result in abnormal fetal growth. The developing fetus is particularly sensitive to toxic chemicals during certain periods, and generally the periods associated with the growth of organic systems or certain cell types. Usually the critical stage of induction of dysfunction in humans is 20-70 days after conception. The effect of chemicals (or drugs) on the reproductive system was tragically demonstrated by the use of sedatives in the 1960s. A pregnant woman was given painkillers as a nausea-preventing drug. This drug has no negative effects on adults but is teratogenic and interferes with fetal development. As a result, children whose mothers used the drug during pregnancy were born with significant abnormalities in the limbs or without arms or legs.

HEALTH AND SAFETY OF WORKERS

Hazardous waste sites and emergency releases of toxic substances have the potential to cause adverse health effects in community populations. However, other persons also are at possible health risk due to exposure to substances in hazardous waste (Table **7a** and **b**). In particular, there are safety and health implications for workers whose occupations place them in direct contact with waste sites or waste management facilities. Included in these categories would be waste site remediation workers and workers who dispose of hazardous wastes, for example, incineration workers.

Substances were assigned the same rank when two (or more) substances received equivalent total point scores (Table **8**). CAS RN= Chemical Abstracts Service Registry Number.

Table 7a. Hazardous Substances in PCs, Recyclability, and Their Potential Adverse Health Effect [53].

Substance	Use/Location	Recyclability (%)	Adverse Health Effects
Aluminum	Structural, conductivity/housing, CRT, PWB, connectors	80	Damage to kidney and central nervous system, skin rashes, skeletal problems, respiratory problems including asthma, linked to Alzheimer's disease

(Table 7a) cont.....

Substance	Use/Location	Recyclability (%)	Adverse Health Effects
Antimony	Diodes/housing, PWB, CRT	0	Pneumoconiosis, heart problems, stomach ulcers
Arsenic	Doping agents in transistors/PWB, light- emitting diodes	0	Allergic reactions, nausea, vomiting, decreased red and white blood cell production, abnormal heart rhythm, (inorganic arsenic is known human carcinogen)
Barium	Vacuum tube/CRT	0	Breathing difficulties, increased blood pressure, swelling of brain, damage to heart, liver, and kidneys
Beryllium	Thermal conductivity/PWB, Connectors	0	Lung damage, allergic reactions, chronic beryllium disease (beryllium is a suspected human carcinogen)
Cadmium	Rechargeable batteries, blue- green phosphor emitter/housing		Pulmonary and kidney damage, bone fragility, (cadmium is a suspected carcinogen)
Chromium	Decorative, hardener/(steel) housing	0	Ulcers, convulsions, liver and kidney damage, asthmatic bronchitis, liver and kidney problems
Copper	Conductivity/CRT, PWB, connectors	90	Chronic exposure can irritate nose, mouth, and eyes and cause headaches, dizziness, nausea, and diarrhea.
Europium	Phosphor activator/PWB	0	Cancer of the liver and bone
Gallium	Semiconductors/PWB	0	Carcinogen in experimental animals
Germanium	Semiconductors/PWB	0	Carcinogen in experimental animals
Gold	Connectivity, conductivity, PWB, connectors	99	

Table 7b. Continued: Hazardous Substances in PCs, Recyclability, and Their Potential Adverse Health Effect [53].

Substance	Use/Location	Recyclability (%)	Adverse Health Effects
Indium	Transistor, rectifiers/PWB	60	Damage to the heart, kidney, liver, and may be teratogenic
Lead	Metal joining, radiation shield, CRT, PWB	5	Damage to central and peripheral nervous system, kidneys, and brain development
Manganese	Structural, magnetivity (steel), housing	0	
Mercury	Batteries, switches/housing, PWB	0	Chronic brain, kidney, lung, and fetal damage
Nickel	Structural, magnetivity (steel), housing, CRT, PWB	80	Allergic reactions, asthma, impaired lung function, chronic bronchitis, carcinogenic
Niobium	Welding alloy/housing	0	May cause kidney damage and scarring of the lung if inhaled
Palladium	Connectivity, conductivity/PWB, Connectors	95	Damage to bone marrow, liver, and kidneys; may also cause skin, eye, and respiratory tract infections
Platinum	Thick film conductor/PWB	95	Cancer, damage to kidneys, intestines, hearing, and bone marrow
Rhodium	Thick film conductor/PWB	50	Carcinogen, central nervous system dysfunction in animals
Ruthenium	Resistive circuit	80	Yet to be determined, may cause skin and eye irritation
Selenium	Rectifiers/PWB	70	Breathing problems, skin, lung and throat irritation, stomach pain
Silver	Connectivity, conductivity/PWB, connectors		Breathing problems, skin, lung and throat irritation, stomach pain
Tantalum	Capacitors/PWB, power supply	0	Eye and skin irritation
Terbium	Green phosphor activator, dopant/CRT, PWB	0	Eye and skin irritation, other effects are yet to be investigated
Tin Titanium	Pigment, alloying agent/housing	0	No conclusive evidence
Vanadium	Red phosphor emitter/CRT	0	Damage to lungs, throat and eyes, possible kidney and liver dysfunction
Yttrium	Red phosphor emitter/CRT	0	Damage to lungs and liver of animals reported

(Table 7b) cont.....

Substance	Use/Location	Recyclability (%)	Adverse Health Effects
Zinc	Battery, phosphor emitter/PWB, CRT	60	Very high levels can damage the pancreas, and it is dangerous for unborn and newborn children

Table 8. Substance Priority List [54].

2017 Rank	Substance Name	Total Points	CAS RN
1	Arsenic	1674	7440-38-2
2	Lead	1531	7439-92-1
3	Mercury	1458	7439-97-6
4	Vinyl chloride	1358	75-01-4
5	Polychlorinated biphenyls	1345	1336-36-3
6	Benzene	1329	71-43-2
7	Cadmium	1320	7440-43-9
8	Benzo(a)pyrene	1306	50-32-8
9	Polycyclic aromatic hydrocarbons	1279	130498-29-2
10	Benzo(b)fluoranthene	1251	205-99-2

BIOLOGICAL IMPACTS ASSOCIATED WITH HAZARDOUS WASTES

There are many consequences from exposure to hazardous waste; this is whatever makes researchers resort to the identification of biological markers that pave the way for the detection of such effects [55]. The term "biomarker" has been used, in general, to mean a signal or indicator that serves as a guide or sets a boundary [56]. A biomarker refers to any indicator of the status of a biological system. Biologic markers don't seem to be new; such blood lead, urinary phenol levels in aromatic hydrocarbon exposure, and liver assays, once solvent exposure have long been used in the activity, and the follow up recent exposures to those compounds [57, 58]. Toxic exposure is one in the potential causes for abnormal biomarker, however, there are several in which an organ system departs from the physiological condition. Malignant tumors, pregnancy, infection, trauma, metabolic errors, and physiological stress can also cause changes in biomarkers [52]. The using of comparison populations, questionnaires, analytical measurements of specific toxicants, and longitudinal follow of exposed populations are required, to see whether or not the presence of a biomarker reflects the consequences of nephrotoxic exposure and predicts future health effects [52, 53] Biomarkers are wont to identify: 1) condition to the nephrotoxic effects of environmental pollutants; 2) exposure to such toxicants; and 3) biological effects of exposure [52, 59]. Condition markers are measurable indicators of biological factors that influence the chance of illness. Exposure markers are measurable indicators that may occur when the body is exposed to toxicant(s). The foremost obvious example of an exposure marker is the measured level of a toxicant in an organism [60]. A biomarker of effect, measured usually in blood or urine, reflects change in an organ resulting from toxic effects, if other causes for the change are ruled out [61, 62].

What distinguishes this generation of research on the markers is that the greater degrees of analytical sensitivity that exist to look for these markers as well as the flexibility these markers offer researchers to elucidate events in the duration between exposure to any harmful substance and therefore clinical symptoms. There are vital areas to respond to levels of accuracy that were unknown 20 years ago [63]. For example, over the past few years, more than 400 proteins have been identified on sperm. Theoretically, the chemicals encountered and already discovered can form in protamine, hemoglobin, and other vital human proteins [64]. Coupled with these developments in the sensitivity of the analysis is that the demand to believe the presence of various factors will affect the appearance of those biomarkers. All individuals with similar exposures do not develop upset, signs of exposure, or illness. Many host and genetic factors are responsible for this difference in responses. Biomarkers may represent signals in the communication or progress of events between causal environmental exposure and the resulting disease [63].

These signs are typically chemical, molecular, genetic, immunologic, and physiological signals of an event. This methodology of the risk assessment will currently be complemented by associating exposure to clinical fatigue which will result in morbidity or mortality a lot of absolutely, a way that identifies overlapping relationships with a lot of accurately or in larger detail than within the past [65]. As a result, health events are less probably to be seen as bilateral phenomena (presence or absence of disease) than they must be seen as a series of changes that hope for continuity through wrong adaptation, feebleness, and death [57].

Some biological markers of exposure like deoxyribonucleic acid or macromolecule adducts may be specific. Approaches of deoxyribonucleic acid, hemoglobin components, and alternative proteins directly indicate the presence of an organic foreign substance, and its interaction with a crucial critical molecule or molecule replacement [64, 66]. Confirmed signs of influence also can be accustomed resolve questions on whether or not a mix of signs and symptoms indicate or doesn't indicate an ailment or an early pathological process. Moreover, recognition of signs of influence will yield timely and prudent interventions [57].

Biomarkers for exposure are sometimes restricted to studies that monitor the health of unsafe waste employees, particularly those exposed to unsafe metals and pesticides, and also the results of such efforts have to date been low. Reflective the evolving nature of this field, the signs of biological impact have conjointly been known in occupational studies and this confirms the evolution of this science and its expansion to cover many areas and serve as much as possible [67].

The traditional liver injury tests provide measures that are not sensitive to total liver function, as they do not reflect specific chemical exposure. Moreover, the liver is a target organ for many of the toxic substances present in waste sites. Modern techniques that detect urinary excretion from metabolites and other specific abnormalities should be validated so that they can be incorporated into biological monitoring [68]. Liver injury in hazardous waste workers is difficult to explain with possible exposure to hepatotoxicity for some reasons including The true predictive value of liver injury tests is unknown, and the longstanding diagnosis of mild elevation in liver function is unknown or weak, and the overlapping causes of nonspecific liver disease (such as viral diseases, alcohol consumption and minor exposure to liver toxicity) are unexplored [68].

GENETICS EFFECTS

In the latest decades, the concern of the genomics and genetics research community has transferred towards comprehending the complex prevalent disorders. Some cases of common disorders are characterized by a strong influence of germline mutations in a single gene; these will be referred to as 'monogenic subtypes. Within a great many cases prevalent ailments possess the multifactorial etiology: that they have been triggered through lots of genes together with environmental agents, involving gene-gene and/or gene-environment interactions [69, 70].

The following technologies are catalytic in supporting the implementation of the connectivity approach to chemical risk assessment: genomics, proteomics, metabonomics, and biology-based dose-response modeling. Genomics, or more befittingly, transcriptomics measures the modulation in the expression of the mRNA to determine disarray at the molecular level due to xenobiotics. Differential gene expression means differentiated coding of proteins, which, in turn, leads to modifications of the protein equilibrium defined as the proteome [55].

Exposure represents the total exposures since pregnancy and beyond, including external and internal exposures and adjustable risk factors leading to the disease and prediction of hazardous substances [71]. Revealing this will help us to understand the complex network of relationships between environmental exposures, lifestyle, genetics and disease. This process includes environmental exposures and genetic variability that are already measured and reliably correlated by mechanical analysis of toxicity pathways (Fig. **1**). To understand the interaction between environmental exposure and disease induction, we need a good knowledge of the biological disorders caused by chronic exposure to environmental stresses and to identify those that overcome the equilibrium barrier and succeed in making changes in the cellular or histological environment and ultimately show the apparent patterns of disease [72].

Although markers of influence based on other things that may be very important, such as cellular genetic effects, have been used in populations close to hazardous waste sites, they have not yet been reported. One impressive exception is that several studies using proteins that show mutated or overexpressed genes in experimental studies of workers in the hazardous waste and foundry industries have shown the predictive benefit of using protein products as a carcinogenic agent and then as possible markers of cancer risk. Abnormal patterns of tumor-related protein products have been detected in both the biological fluids of individuals with cancer and in individuals who subsequently develop a pre-tumor colonic tumor. At the time of the evaluation of tumor-related proteins, all workers surveyed were clinically healthy. On the other hand, a person who has been exposed in his workplace to Polychlorinated biphenyls, asbestos, and pesticides, as well as smoking 20 cigarettes a day, has developed an abnormal Ras oncogene that induced a colon tumor [73 - 75].

Oncogene Ras can be activated by two mechanisms: first, point mutation and the second is an over-expression of the progeny of the tumor gene. Researchers hypothesize that exposure to asbestos, in this case, may have led to gene activation by the second method; asbestos fibers can transfect the outer segments of deoxyribonucleic DNA, such as carcinogens and promoter sequences. But after the removal of the colon tumor, the encoded p21 protein was not detected in the patient's serum. This condition indicates the potential benefit of tumor gene protein products as signs of early detection of cancer [74].

Although the risk of cancer provides a major focus in many types of research on biomarkers that can detect early and accurate, the risk of human reproduction provides another focal point, associated with much shorter intervals between exposure to cancer-inducing substances and evidence of a health effect relevant [75]. Several studies have revealed that workplace exposure to males affects their reproductive capacity, as well as the health of their offspring. When it was found that a group of employees who had been exposed to dipromochloropropane were unable to have children [76, 77]. The researchers confirmed that the exposed workers had abnormal semen efficiency analyzes in terms of the morphometric results of the testes or sperms and teratogenic motility [78]. Previous studies have also revealed similar effects in multiple species, including testicular atrophy and clear evidence that the hormone system is affected by these toxic substances and hazardous wastes such as organophosphorus. Although the risk of cancer provides a major focus on many kinds of research on markers, the risk of human reproduction provides another focal point, which relates to much shorter periods between exposure and evidence of a relevant health effect [79 - 81]. Such materials caused human congenital abnormalities; Human exposure to pesticides has also been associated with epigenetic transformations and the emergence of chronic diseases. Besides, the main risk factors for these hazardous substances include genetic predisposition, taking precautions in reducing folic acid intake and environmental factors such as anticonvulsants, maternal obesity, and maternal diabetes. Inherited germline mutations can have devastating effects on offspring affected by toxic substances to parents and have significant socio-economic repercussions at the population level. Animal studies have shown that exposure to hazardous substances can increase the risk of multiple mutations in germ cells [82, 83]. Exposure to toxic substances leads to a host of adverse effects on biological systems. The general agreement is needed on any of these critical effects that point to some aspects of the disease response. Therefore, it is necessary to link the critical effects to the dose estimates, to determine the factors that affect the dose and to determine the level or degree of exposure to distinguish the toxic dose from others [84 - 86].

CYTOTOXIC MECHANISMS OF HAZARDOUS WASTES

Cytotoxic waste is highly hazardous and should never be landfilled or discharged into the sewerage system [87].

It is important to know the mechanisms through which cytotoxic substances permeate through the environment, and attention to hazardous waste where cytotoxic waste is considered extremely hazardous and should not be discharged into the sewage system. It also requires the complete destruction of all cytotoxic substances at burning temperatures up to 1200°C. Burning at low temperatures may release toxic vapors into the atmosphere that negatively affect cells. Incineration in most single-chamber incinerators or by open-air incineration is also a risk for the disposal of cytotoxic waste [87]. Chemical degradation methods require expertise and are customized for drug residues and for cleaning contaminated urinals, spillages, and protective clothing [88].

The ability to produce highly reactive chemical entities, such as free radicals, is a serious connotation of heavy metals and toxic waste. Key factors that make the greatest contribution to the toxicity of these different minerals include reactive oxygen species (ROS) and nitrogen types (RNS) that impede oxidation and reduction systems [25]. ROS that are distinguished by their high chemical reactivity, include free radicals like superoxide

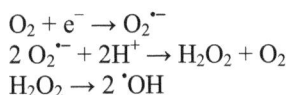

$$O_2 + e^- \rightarrow O_2^{\cdot-}$$
$$2\,O_2^{\cdot-} + 2H^+ \rightarrow H_2O_2 + O_2$$
$$H_2O_2 \rightarrow 2\,{}^\cdot OH$$

Intracellular generation of superoxide anion $(O_2^{\cdot-})$ primarily occurs non-enzymatically through the intervention of redox components such as semi-ubiquinone (a component of the mitochondrial electron transport chain) [89], or *via* the intervention of enzymes such as NADPH-oxidase (NOX) [90], xanthine-oxidase or auto-oxidation reactions [91]. Superoxide anion $(O_2^{\cdot-})$ acts as a mild reactant under physiological conditions, with poor ability to cross the biological membranes. Upon interaction with nitric oxide (NO), production of peroxynitrite $(ONOO^-)$ transforms superoxide into very reactive intermediates such as hydroxyl radical $({}^\cdot OH)$, which have a very short half-life [92].

$$NO + O_2^{\cdot-} \rightarrow ONOO^- + H^+ \rightarrow {}^\cdot OH + {}^\cdot NO_2$$

Through the involvement of nitric oxide synthase isozymes like endothelial nitric oxide synthase and (eNOS) mitochondrial nitric oxide synthase (mtNOS), generation of nitric oxide occurs *via* conversion of L-arginine to citrulline. $NO^\cdot$ has been shown to have greater stability in oxygen deprived environments. Because of its amphipathic nature, $NO^\cdot$ easily diffuses through the cytoplasm and plasma membranes. Upon interacting with superoxide anion, NO generates peroxynitrite $(ONOO^-)$ [93].

An increase in ROS/RNS production or decrease in ROS-scavenging activity that arises as a result of exogenous stimuli has been found to alter cellular functions through direct modifications of biomolecules and/or by aberrant stimulation/suppression of certain signalling pathways affecting growth factor receptors.

THE REMEDIATION OF UNDESIRABLE EFFECTS OF HAZARDOUS WASTES ON HUMAN HEALTH

Antioxidant enzymes are the body's first line of defense against free radicals and maintain of the cellular homeostasis. These antioxidants, such as (GPx) glutathione peroxidase, GSH-reductase, GSH-transferase, catalase (CAT) and superoxide dismutase (SOD) that work to sweep the residues of mono oxygen and oxidized or reductive form where these enzymes play an effective role in protecting the body from the destructive effect of free radicals [94]. Additionally, the interference of minerals in the synthesis of these enzymes such as manganese, zinc, copper to superoxide enzyme. This enzyme is based on copper, zinc and manganese and is found in plants and animals, with a high concentration in the brain, liver, heart, red blood cells and kidneys [95, 96].

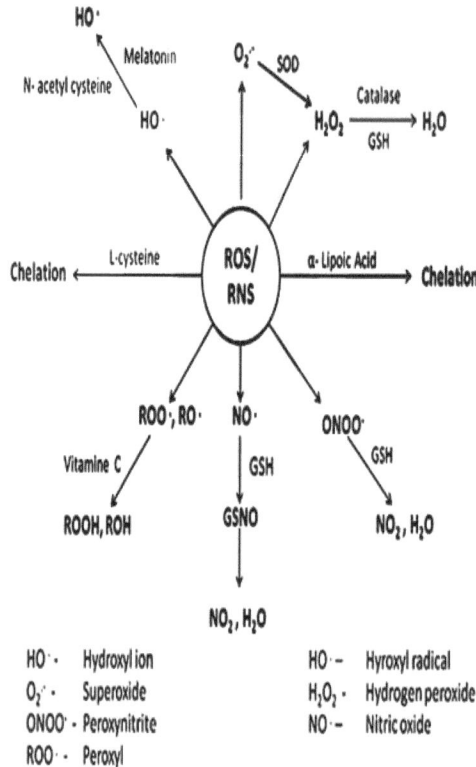

HO·

Melatonin $O_2^{\cdot-}$ SOD

N- acetyl cysteine

HO·   Catalase

H_2O_2 ——→ H_2O
GSH

Chelation ← L-cysteine   ROS/   α- Lipoic Acid → **Chelation**
RNS

ROO·, RO·   NO·   ONOO·

Vitamine C   GSH   GSH

ROOH, ROH   GSNO NO_2, H_2O

NO_2, H_2O

HO··	Hydroxyl ion	HO·−	Hyroxyl radical
$O_2^{\cdot}$·	Superoxide	H_2O_2·	Hydrogen peroxide
ONOO··	Peroxynitrite	NO·−	Nitric oxide
ROO··	Peroxyl		

Fig. (3). Modes of action of different antioxidants in mitigating the toxic effects imposed by metals [25].

Maintaining the necessary and controlled distribution of toxic metal ions is an effective means of protecting against the harmful effects of heavy metals (Jan *et al.* 2015). Accordingly, the enhanced understanding of beneficial antioxidant defense mechanisms supports their role in the treatment of oxidative stress caused by these metals and the hazardous wastes that result from them [97, 98]. Studies on the molecular mechanisms involved in eliminating toxicity from heavy metals are necessary. Besides, detailed mechanism studies underpinning the beneficial effects of dietary antioxidants for optimal dose and duration of treatment will be useful in the development of harmonic strategies as part of effective treatment regimens to improve clinical healing in cases of toxic effects of hazardous substances (Fig. **3**) [25, 99].

Hazardous waste can be disposed of in place or another location through a variety of processes depending on the type of waste and the specific situation. However, the residual waste no longer possesses toxic properties [100]. Detoxification is strongly recommended at the site of origin of hazardous waste, not only that toxic waste producers are more able to detoxify their waste as they originate, but they also protect final disposers from accidental spills or breakdown systems [101]. One of the ways of disposal of solid waste by dumping in landfills, or pits to be prepared for this purpose, lining drilling mud and plastic to ensure that the liquids from the waste does not reach the soil, and collecting leachate, and treated with chemicals so as not to contaminate water sources, Waste such as methane and carbon dioxide are decomposed and released into the atmosphere. When these craters are filled, they are covered with a layer of soil and mud to prevent rainwater from reaching them [101].

Recycling: the reuse of waste to produce new materials, the advantage of this method is that it reduces the need for new resources, and the energy required to recycle materials is less than the energy required to produce a product using new materials, and most of all that Recycling reduces the amount of waste that needs to be disposed of by incineration or landfill. The most important materials that can be recycled, metal, glass, paper, and plastic.

Biogas technology: When wastes containing organic matter are decomposed by anaerobic bacteria, they produce biogas consisting of methane and carbon dioxide. Convert waste to organic fertilizers: This method is based on the collection of residues of kitchen waste, and left open to begin aerobic bacteria, insects, worms, and fungi analysis of organic materials in them, taking care to turn the waste from time to time to allow oxygen to permeate between the waste so that the bacteria do not resort to anaerobic decomposition Which produces methane, and other gases that cause an unpleasant odor. When the process of decomposition is finished, the waste turns into a bio-fertilizer - sometimes called black gold - that can be mixed with soil, or placed around plants [102].

GREEN CHEMISTRY AS A NEW TREND OF REMEDIATION

Green chemistry is a method designed to eliminate the use and generation of hazardous materials by designing better manufacturing processes for chemical products. This is a revolution in the discovery of specifically treated drugs; the determined aim is to make the end product and byproducts less toxic to guide decisions made during chemical production [103]. The aspects of the manufacturing process that have been considered include the initial selection of chemicals, the chemical synthesis mechanism, the final products of the process, and the management of toxic products produced during production. By reducing intrinsic risks to chemical products, the risk from that product is reduced (Fig. **4**) [104].

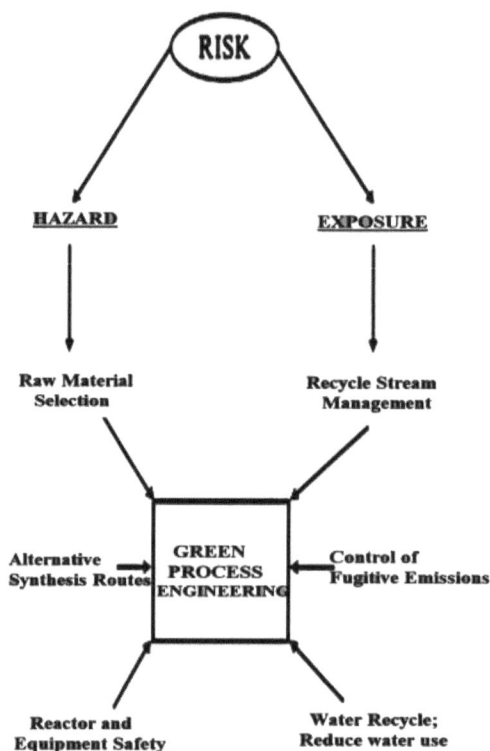

Fig. (4). Green processes to reduce the risk of chemical manufacturing [103].

Green science standards were initially set up well by Anastas and Warner [105]. So far it has been applied to the plan of a wide range of elements and techniques with points to reduce the risk of chemicals and hazardous waste and dependence on nature, thus avoiding pollution. The use of these standards in nanoscience will support the separation of the production and handling of safer nanomaterials and tools [106, 107]. Green nanotechnology contains the use of green science standards in the nano elements plan, the extension of green nanomaterial generation techniques, and the use of nature in the manufacture of green nanomaterials [108]. This technique aims to change the understanding of the properties of nanomaterials, including their specific properties, as well as the structure of nanomaterials that can be combined into superior elements that are safer for human and environmental well-being. The use of plant extracts to incorporate them into nanomaterial synthesis is conceivable. It is an interesting option for obtaining nanoparticle molecules using all environmentally friendly procedures that maintain their ideal properties for natural functions [108].

CONCLUSIONS

Without a doubt, hazardous waste is toxic to many of the animal species present at the waste sites. Several studies have indicated that many sites have not been adequately evaluated for the content of toxic substances, or possible methods of human exposure. Human health studies also provide evidence that serious health impacts cannot be ruled out. It is of the utmost importance to know whether hazardous waste programs to develop successful strategies for cleaning hazardous waste sites. These measures require information on the range of potential and actual human exposure to hazardous wastes, and on the possible health effects that may be associated with vulnerabilities. With genotoxic, carcinogenic agents, which are likely to be a principal cause of concern at low exposure levels, the initial adverse effect is thought to be the induction of double-strand breaks in the DNA of stem cells or their immediate progenitors. It is becoming feasible to culture such cells *in vitro* and this may be a promising approach to assessing the impacts of such agents either singly or in combination, *e.g.* by studying the induction of mutations, chromosomal aberrations, genomic instability or other sequelae of DNA misrepair. However, this addresses only the initial induction of effects at the sub-cellular level. Additional modelling, supported by data, is required to interpret these results in terms of likely increases in cancer induction.

Tumor initiation, proliferation and progression all need to be addressed. Furthermore, for both exposure to ionizing radiations and chemical pollutants, standards for environmental protection are generally based on precautionary approaches. These include definitions of compliance values set by reference to the sensitivity of the most sensitive species, use of precautionary uncertainty factors, and use of cautious over-estimates in exposure calculations. In some contexts, notably with ionizing radiations, the compliance values are described as screening thresholds, *i.e.* exposures above the thresholds are an indication of a need for further investigation, but do not necessarily imply that adverse effects will be observed. Treatment of the effects of those hazardous wastes requires further research to obtain treatments that have no side effects on human health. This has led scientists to resort to green chemistry, which has revolutionized the therapeutic and hope to achieve the difficult equation in the activation of the defenses of the body against the effects of these hazardous waste without unwanted side effects.

CONSENT FOR PUBLICATION

Not applicable.

CONFLICT OF INTEREST

The author declares that there is no conflict of interest in this chapter.

ACKNOWLEDGEMENTS

Declared none.

REFERENCES

[1]   Fazzo L, Minichilli F, Santoro M, *et al.* Hazardous waste and health impact: a systematic review of the scientific literature. Environ Health 2017; 16: 107.

[2]   Dongo K, Tiembré I, Koné Blaise A, *et al.* Exposure to toxic waste containing high concentrations of hydrogen sulphide illegally dumped in Abidjan, Cote d'Ivoire. Environ Sci Pollut Res Int 2012; 19(8): 3192-9.

[3]   Hallenbeck WH. Quantitative Risk Assessment for Environmental and Occupational health. Boca Raton: Lewis Publishers 1993.

[4]   Faustman EM, Omenn GS. Risk Assessment. Casarett and Doull's Toxicology: The Basic Science of Poisons. New York: McGraw-Hill 1996; pp. 75-88.

[5]   Perera F, Mayer J, Santella RM, *et al.* Biologic markers in risk assessment for environmental carcinogens. Environ Health Perspect 1991; 90: 247-54.

[6]   Schleyer TK, Teasley SD, Bhatnagar R. Comparative Case Study of Two Biomedical Research Collaboratories. Los Angeles, California, USA 2001; pp. 166-9.

[7]   Triassi M, Alfano R, Illario M, Nardone A, Caporale O, Montuori P. Environmental Pollution from Illegal Waste Disposal and Health Effects: A Review on the Triangle of Death. Int J Environ Res Public Health 2015; 12: 1216-36.

[8]   Krishna IM, Manickam V. Environmental Management: Science and Engineering for Industry. Butterworth-Heinemann; 2017 Jan 23.

[9]   Krewski D, Acosta D, Andersen M, *et al.* Toxicity testing in the 21st century: a vision and a strategy. J Toxicol Environ Health, Part B 2010; 13(2-4): 51-138.

[10]   Azam AG, Zanjani BR, Balali-Mood M. Effects of air pollution on human health and practical measures for prevention in Iran. J Res Med Sci 2016; 21: 65-76.

[11]   Yamamoto SS, Phalkey R, Malik AA. A systematic review of air pollution as a risk factor for cardiovascular disease in South Asia: limited evidence from India and Pakistan. Int J Hyg Environ Health 2014; 217(2-3): 133-44.

[12]   Assi MA, Hezmee MNM, Haron A, Sabri MYM, Rajion MA. The detrimental effects of lead on human and animal health. Vet World 2016; 9(6): 660-71.

[13]   Tsatsakis AM, Vassilopoulou L, Kovatsi C, *et al.* The dose response principle from philosophy to modern toxicology: The impact of ancient philosophy and medicine in modern toxicology science. Toxicol Rep 2018; 5: 1107-13.

[14]   Wheeler DM, Wheeler MM. Stereoselective Syntheses of Doxorubicin and Related Compounds. In: Studies in Natural Products Chemistry. Elsevier 1994; Vol. 14: pp. 3-46.

[15]   Hernández AF, Parrón T, Tsatsakis AM, Requena M, Alarcón R, López-Guarnido O. Toxic effects of pesticide mixtures at a molecular level: their relevance to human health. Toxicology 2013; 307: 136-45.

[16]   Barlow S. Threshold of toxicological concern (ttc) a tool for assessing substances of unknown toxicity present at low levels in the diet. Belgium: International Life Sciences Institute 2005.

[17]   Rowe PH. Statistical methods for categorised endpoints in *in silico* toxicology. *In Silico* Toxicology: Principles and Applications. In: Cronin MTD, Madden JC, Eds. Cambridge, UK: The Royal Society of Chemistry 2010; pp. 252-74.

[18]   Hodgson E. A Textbook of Modern Toxicology. 3rd ed., Wiley-Interscience 2004.

[19]   Adeola FO. Hazardous Wastes , Industrial Disasters, and Environmental Health Risks Local and Global Environmental Struggles175 Fifth Avenue, New York, NY 10010.: Palgrave Macmillan® in the United States- a division of St. Martin's Press LLC 2011; p. 39.

[20]   Doull J, Klaassen CD, Amdur MO. Casarett and Doull's Toxicology: The Basic Science of Poisons. New York, NY: Macmillan Publishing Co., Inc. 1986.

[21]   Maechel and Ash I. Handbook of Green Chemicals. Technology & Engineering. 2nd ed. Resources, USA: Synapse Info 2004; p. 1345.

[22]   Jeng S, Gupta N, Wrensch M, *et al.* Prevalence of congenital hydrocephalus in California, 1991–2000. Pediatr Neurol 2011; 45(2): 67-71.

[23]   EPA (Environmental Protection Agency). 2017.https://www.epa.gov/hw/learn-basics-hazardous-waste

[24]   Ashraf MA, Maah MJ, Yusoff I. Soil Contamination. Risk Assessment 2014; pp. 1-56.

[25]   Jan AT, Azam M, Siddiqui K, Ali A, Choi I, QMR. Heavy Metals and Human Health: Mechanistic Insight into Toxicity and Counter Defense System of Antioxidants. Int J Mol Sci 2015; 16(12): 29592-630.

[26]   International Agency for Research on Cancer. List of classification by cancer sites with sufficient or limited evidence in humans International Agency for Research on Cancer 2016; Volumes 1 to 117

[27]   Altekruse SF, McGlynn KA, Reichman ME. Hepatocellular carcinoma incidence, mortality, and survival trends in the United States from 1975 to 2005. J Clin Oncol 2009; 27(9): 1485.

[28]   Mastrangelo G, Fedeli U, Fadda E, *et al.* Increased risk of hepatocellular carcinoma and liver cirrhosis in vinyl chloride workers: synergistic effetcs of occupational exposure with alcohol intake. Environ Health Perspect 2004; 112(11): 1188-92.

[29]   Johnson D, Groopman JD. Toxic liver disorders. Environmental and Occupational Medicine. Philadelphia: Lippincott Williams & Wilkins 2007; pp. 789-99.

[30]   Schaider LA, Senn DB, Brabander DJ, Mccarthy KD, Shine JP. Characterization of zinc, lead, and cadmium in mine waste: Implications for transport, exposure, and bioavailability. Environ Sci Technol 2007; 41: 4164-71.

[31]   Tsuda A, Henry FS, Butler JP. Particle transport and deposition: basic physics of particle kinetics. Compr Physiol 2013; 3(4): 1437-71.

[32]   Williams PL, Brooks JD, Inman AO, Monteiro-Riviere NA, Riviere JE. Determination of physicochemical properties of phenol, p-nitrophenol, acetone and ethanol relevant to quantitating their percutaneous absorption in porcine skin. Res Commun Chem Pathol Pharmacol 1994; 83(1): 61-75.

[33]   Bennett E, Peterson G, Gordon L. Understanding relationships among multiple ecosystem services. Ecol Lett 2009; 12: 1394-404.

[34]   Nebel BJ, Wright RR. Environmental Science: The Way the World Works. Upper Saddle River, NJ: Prentice Hall 2000.

[35]   Risher JF, De Rosa CT, Jones DE, Murray HE. Letter to the editor: updated toxicological profile for mercury. Toxicol Indus Health 1999;15(5):480-516.

[36]   Nriagu JO. History of global metal production. Science 1996; 22: 223-4.

[37]   Adeola FO. Taxonomy of Hazardous Wastes. In: Hazardous Wastes, Industrial Disasters, and Environmental Health Risks. Palgrave Macmillan: New York 2011; pp. 25-53).

[38]   Rice D, Silbergeld E. Lead Neurotoxicity: Concordance of Human and Animal Research.Toxicology of Metals. New York: CRC- Lewis Publishers 1996; pp. 659-75.

[39]   2012.https://www.epa.gov/sites/production/files/2015-05/documents/402-k-10-008.pdf

[40]   Saleh HM. Treatment and Solidification of Hazardous Organic Wastes: Radioactive Cellulose-based Wastes. Germany: LAP Lambert Academic 2012.

[41]   de Groot PM, Wu CC, Carter BW, Munden RF. The epidemiology of lung cancer. Transl Lung Cancer Res 2018; 7(3): 220-33.

[42]   Wigle DT, Lanphear BP. Human health risks from low-level environmental exposures: No apparent safety thresholds. PLoS Med 2005; 2(12): e350.

[43]   Pasetto R, Ranzi A, De Togni A, *et al.* Cohort study of residents of a district with soil and groundwater industrial waste contamination. Ann Ist Super Sanita 2013; 49(4): 354-7.

[44]   Fantini F, Porta D, Fano V, *et al.* Epidemiologic studies on the health status of the population living in the Sacco River Valley. Epidemiol Prev 2012; 36(5) (Suppl. 4): 44-52.

[45]   Fazzo L, Belli S, Minichilli F, *et al.* Cluster analysis of mortality and malformations in the Provinces of Naples and Caserta (Campania Region). Ann Ist Super Sanita 2008; 44(1): 99-111.

[46]   Martuzzi M, Mitis F, Bianchi F. Cancer mortality and congenital anomalies in a region of Italy with intense environmental pressure due to waste. Occup Environ Med 2009; 66(11): 725-32.

[47]   Giovannini A, Rivezzi G, Carideo P, *et al.* Dioxins levels in breast milk of women living in Caserta and Naples: assessment of environmental risk factors. Chemosphere 2014; 94: 76-84.

[48]   Porta D, Milani S, Lazzarino AI, Perucci CA, Forastiere F. Systematic review of epidemiological studies on health effects associated with management of solid waste. Environ Health 2009; 8: 60-74.

[49]   Vrijheid M. Health effects of residence near hazardous waste landfill sites: a review of epidemiologic literature. Environ Health Perspect 2000; 108 (Suppl. 1): 101-12.

[50]   Guérard M, Marchand C, Funk J, Christen F, Winter M, Zeller A. DNA damage response of 4-Chlor--Ortho-Toluidine in various rat tissues. Toxicol Sci 2018; 163(2): 516-24.

[51]   Hassan AI, Alam SS. Evaluation of mesenchymal stem cells in treatment of infertility in male rats. Stem Cell Res Ther 2014; 5: 131.
[http://dx.doi.org/10.1186/scrt521]

[52]   Apostoli P, Porru S, Bisanti L. Critical aspects of male fertility in the assessment of exposure to lead. Scand J Work Environ Health 1999; 25: 40-3.

[53]   Agency for Toxic Substances and Disease Registry (ATSDR). http://www.atsdr.cdc.gov/toxprofiles/phs54.html1996.

[54]   Agency for Toxic Substances and Disease Registry (ATSDR). Page last reviewed 2017.https://www.atsdr.cdc.gov/SPL/

[55]   Sarigiannis DA. Assessing the impact of hazardous waste on children's health: The exposome Paradigm. Environ Res 2017; 158: 531-41.

[56]   Straight M, Amler RW, Vogt RF, Kipen HM. Biomarker testing for the assessment of populations exposed to hazardous chemicals. InMethods of Pesticide Exposure Assessment.
Springer, Boston, MA 1995: pp. 165-175.

[57]   NRC. National Research Council; Committee on Environmental Epidemiology Board on Environmental Studies and Toxicology Commission on Life Sciences. D.C.: Public Health and Hazardous Wastes. National Academy Press Washington 1991; Vol. 1: p. 327.

[58]   Fougère B, Vellas B, van Kan GA, Cesari M. Identification of biological markers for better characterization of older subjects with physical frailty and sarcopenia. Transl Neurosci 2015; 6(1): 103-10.

[59]   Curry PB, Iyengar S, Maloney PA, Maroni M. Methods of Pesticide Exposure Assessment. 166.Springer Science & Business Media 2013; p. Nov 21

[60]   Silins I, Högberg J. Combined toxic exposures and human health: Biomarkers of exposure and effect. Int J Environ Res Public Health 2011; 8(3): 629-47.

[61]   Watson WP, Mutti A. Role of biomarkers in monitoring exposures to chemicals: present position, future prospects. Biomarkers 2004; 9(3): 211-42.

[62]   Costa LG. Central nervous system toxicity biomarkers. In: Biomarkers in Toxicology. Academic Press 2019; pp. 173-185.

[63]   Schulte PA, Ringen K. Notification of workers at high risk: an emerging public health problem. Am J Public Health 1984; 74(5): 485-91.

[64]   NRC (National Research Council). Biologic markers in environmental health research. Environ Health Perspect 1987; 74: 3-9.

[65]   Oliveira MB. Technology and basic science: the linear model of innovation. Scientiae Studia 2014;12(SPE):129-46.

[66]   Yun BH, Guo J, Bellamri M, Turesky R. DNA adducts: Formation, biological effects, and new biospecimens for mass spectrometric measurements in humans. Mass Spectrom Rev 2018; 11 [Epub ahead of print].
[http://dx.doi.org/10.1002/mas.21570]

[67]   Gochfield M. Biological monitoring of hazardous waste workers. Metals Occup Med 1990; 5: 25-31.

[68]   Hodgson MJ, Goodman-Klein BM, van Thiel DH. Evaluating the liver in hazardous waste workers. Occup Med 1990; 5: 67-78.

[69]   Becker F. Genetic testing and common disorders in a public health framework: how to assess relevance and possibilities. Eur J Hum Genet 2011; 19 (Suppl. 1): S6-S44.

[70]   Klimek P, Aichberger S, Stefan Thurner S. Disentangling genetic and environmental risk factors for individual diseases from multiplex comorbidity networks. Sci Rep 2016; 6: 1-10.

[71]   Wild CP. Complementing the genome with an "exposome": the outstanding challenge of environmental exposure measurement in molecular epidemiology. Cancer Epidemiol Biomarkers Prev 2005; 14: 1847-50.

[72]   Boekelheide K, Blumberg B, Chapin RE, *et al.* Predicting later-life outcomes of early-life exposures. Environ Health Perspect 2012; 120(10): 1353-61.

[73]   Rowland I, Gibson G, Heinken A, *et al.* Gut microbiota functions: metabolism of nutrients and other food components. Eur J Nutr 2018; 57(1): 1-24.

[74]   Cortessis VK, Thomas DC, Levine AJ, *et al.* Environmental epigenetics: prospects for studying epigenetic mediation of exposure–response relationships. Hum Genet 2012; 131(10): 1565-89.

[75]   Brandt-Rauf PW, Niman HL. Serum screening for oncogene proteins in workers exposed to PCBs. Br J Ind Med 1988; 45: 689-93.

[76]   Brandt-Rauf PW, Niman HL, Smith SJ. Correlation between serum oncogene protein expression and the development of neoplastic disease in a worker exposed to carcinogens. J R Soc Med 1990; 83: 594-5. a

[77]   Nas/Nrc. Report of the oversight committee. Biologic Markers in Reproductive Toxicology 1989.

[78]   Whorton D, Krauss RM, Marshall S, Milby TH. Infertility in male pesticide workers. Lancet 1977; II: 1259-61.

[79]   Babich H, Davis DL, Stotzky G. Dibromochloropropane (DBCP): A review. Sci Total Environ 1981; 17: 207-21.

[80]   Torkelson TR, Sadek SE, Rowe VK, *et al.* Toxicologic investigations of 1,2-dibromo-3-chloropropane. Toxicol Appl Pharmacol 1961; 3: 545-59.

[81]   Perry MJ, Venners SA, Chen X, *et al.* Organophosphorous pesticide exposures and sperm quality. Reprod Toxicol 2011; 31(1): 75-9.

[82]   Mostafalou S, Abdollahi M. Pesticides and human chronic diseases: evidences, mechanisms, and perspectives. Toxicol Appl Pharmacol 2013; 268(2): 157-77.

[83]   Kalliora C, Mamoulakis C, Vasilopoulos E, *et al.* Association of pesticide exposure with human congenital abnormalities. Toxicol Appl Pharmacol 2018; 346: 58-75.

[84]   Copp AJ, Stanier P, Greene ND. Neural tube defects: recent advances, unsolved questions, and controversies. Lancet Neurol 2013; 12(8): 799-810.

[85]   DeMarini DM. Declaring the existence of human germ-cell mutagens. Environ Mol Mutagen 2012; 53: 166-72.

[86]   Crump KS, Chen C, Louis TA. The future use of *in vitro* data in risk assessment to set human exposure standards: Challenging problems and familiar solutions. Environ Health Perspect 2010; 118(10): 1350-4.

[87]   Chartier Y, Emmanuel J, Pieper U, Prüss A, Rushbrook P, Stringer R, Eds. Management of Wastes from Health-Care Activities' WHO Blue Book. 2nd ed., Malta: World Health Organization 2014.

[88]   Easty AC, Coakley N, Cheng R, *et al.* Safe handling of cytotoxics: Guideline recommendations. Curr Oncol 2015; 22: e27-37.

[89]   Flora G, Gupta D, Tiwari A. Toxicity of lead: A review with recent updates. Interdiscip Toxicol 2012; 5: 47-58.

[90]   Pagliaro P, Femminò S, Penna C. Redox Aspects of Myocardial Ischemia/Reperfusion Injury and Cardioprotection.Oxidative Stress in Heart Diseases. Singapore: Springer 2019; pp. 289-324.

[91]   Cadenas E, Davies KJ. Mitochondrial free radical generation, oxidative stress and aging. Free Radic Biol Med 2000; 29: 222-30.

[92]   Pastor N, Weinstein H, Jamison E, Brenowitz M. A detailed interpretation of OH radical footprints in a TBP–DNA complex reveals the role of dynamics in the mechanism of sequence specific binding. J Mol Biol 2000; 304: 55-68.

[93]   Szabo C, Ischiropoulos H, Radi R. Peroxynitrite: Biochemistry, pathophysiology and development of therapeutics. Nat Rev Drug Discov 2007; 6: 662-80.

[94]   Birben E, Sahiner UM, Sackesen C, Erzurum S, Kalayci O. Oxidative stress and antioxidant defense. World Allergy Organ J 2012; 5(1): 9-19.

[95]   Dröge W. Free radicals in the physiological control of cell function. Physiol Rev 2002; 82(1): 47-95.

[96]   Nita M, Grzybowski A. The role of the reactive oxygen species and oxidative stress in the pathomechanism of the age-related ocular diseases and other pathologies of the anterior and posterior eye segments in adults. Oxid Med Cell Longev 2016; 2016: 3164734.

[97]   Poljšak B, Fink R. The protective role of antioxidants in the defence against ROS/RNS-Mediated environmental pollution. Oxid Med Cell Long 2014; 2014: 22.
        [http://dx.doi.org/http://dx.doi.org/10.1155/2014/671539] [PMID: 671539]

[98]   Sorriento D, De Luca N, Trimarco B, Iaccarino G. The antioxidant therapy: New insights in the treatment of hypertension. Front Physiol 2018; 9: 258.

[99]   Unsal V. Natural phytotherapeutic antioxidants in the treatment of mercury intoxication-A review. Adv Pharm Bull 2018; 8(3): 365-76.

[100]   Nemerow NL. Neutralization. InIndustrial Waste Treatment. Elsevier 2007; pp. 35-44.

[101] Panahi A, Shomali A, Sabour MH, Ghafar-Zadeh E. Molecular dynamics simulation of electric field driven water and heavy metals transport through fluorinated carbon nanotubes. J Mol Liq 2019;278:658-71..

[102] Becker TJ, Fullbright N, Robinson R, Terraso D, Sanders JM. Research Horizons [Volume 24, Number 3, Summer 2007].

[103] Ramachandran PA, Shonnard D, Hesketh R, *et al.* Green Engineering: Integration of Green Chemistry, Pollution Prevention, Risk-Based Considerations, and Life Cycle Analysis. Springer International Publishing 1923.
[http://dx.doi.org/https://10.1007/978-3-319-52287]

[104] Poliakoff MJ, Fitzpatrick MJM, Farren TR, Anastas PT. Green chemistry: Science and politics of change. Science 2002; 297(5582): 807-10.

[105] Anastas PT, Warner J. Green Chemistry: Theory and Practice. 1st ed., Oxford, U.K.: Oxford University Press 2000.

[106] Anastas PT, Kirchhoff MM. Origins, current status, and future challenges of green chemistry. Acc Chem Res 2002; 35: 686-94.

[107] Mulvihill MJ, Beach ES, Zimmerman JB, Anastas PT. Green chemistry and green engineering: a framework for sustainable technology development. Annu Rev Environ Resour 2011; 36: 271-93.

[108] Nath D, Banerjee P, Das B. 'Green nanomaterial'-how green they are as biotherapeutic tool. J Nanomed Biother Discov 2014; 4: 1-11.

CHAPTER 5

Heavy Metal Pollution: Sources, Effects, and Control Methods

Sunil Jayant Kulkarni*

Gharda Institute of Technology, Lavel, Maharashtra, India

Abstract: The presence of heavy metals in aquatic systems and atmosphere is attributed to metal plating, chemical synthesis, catalysis, battery, fertilizer, paint, paper, and mining industries. Heavy metals enter the soil through anthropogenic activities and industrial discharges. They enter the human body through plants and vegetables. Heavy metals can cause various acute and chronic diseases in human beings. Heavy metal consumption by fishes and other aquatic animals affect aquatic life and also people consuming them as food. Reduction of the developmental growth, deformities, and increase of developmental anomalies are some of the toxic effects of heavy metals on fishes and aquatic invertebrates. Removal of heavy metals from wastewater is becoming an increasingly important aspect of wastewater treatment. Physico-chemical methods such as filtration and adsorption are simple and effective. These methods can be used to remove heavy metals from wastewater when they are present in very high concentration. Removal method can be selected depending upon the influent concentration, acceptable limit, and amount of effluent. Chemical treatment methods like coagulation, chlorination, flocculation, and precipitation can be used for the treatment of wastewater containing colloids and metal ions due to their simplicity and robustness. Biological treatment methods like advanced oxidation, trickling filters, aeration, activated sludge, aerobic. and anaerobic digestion can be used for heavy metal treatment.

Keywords: Control methods, Heavy metals, Health effects, Pollution, Sources.

INTRODUCTION

Human activities such as smelting, use of insecticides, and burning of fossil fuel lead to severe environmental pollution. Many gases are emitted into the atmosphere as a result of the oxidation of metals present in the fuels. In aquatic systems, the presence of heavy metals is attributed to metal plating, chemical synthesis, catalysis, battery, fertilizer, paint, paper, and mining industries. Heavy metals enter the soil through anthropogenic activities and industrial discharges.

*** Corresponding author Sunil Jayant Kulkarni:** Gharda Institute of Technology, Lavel, Maharashtra, India;
E-mail: suniljayantkulkarni@gmail.com

Gabriella Marfe & Carla Di Stefano (Eds.)

They enter the human body and other animals through plants and vegetables [1 - 4]. Magnetic susceptibility measurements are an indicator of soil pollution [5 - 7]. It is the degree of magnetization of material present in the applied electric field. Unusual susceptibility values can be identified and analysed to quantify soil pollution due to heavy metals [8 - 10]. Many investigations have been carried to study the variation in soil properties because of heavy metal pollution. The presence of heavy metal affects the land use pattern and crop quality [11 - 14]. Efforts are being taken to evaluate and quantify the effect of heavy metal on soil properties [15 - 17]. Predictive and statistical tools are useful in studying soil pollution and predicting the effect of various pollutants on the environment [17 - 20].

Heavy metals have the tendency to accumulate in biological systems, given that they are essential for biological activities. Adverse effects caused by some heavy metals make them important candidate for investigation for their removal from the environment [21 - 24]. Heavy metals like cadmium, nickel, lead, copper, and chromium are used for various applications. Various physical, chemical, physico-chemical, and biological treatment methods can be used for the treatment of wastewater for heavy metal removal [25 - 28]. Many of these methods have limitations in terms of cost-effectiveness, efficiency, environmental safety, and solid waste generation [29, 30]. To explore an efficient, economical, and sustainable alternative for heavy metal removal, many investigators have investigated adsorbents derived from various waste materials, coagulants, flocculating agents, membranes, and ion exchange resins. This book chapter sheds light on studies on heavy metal sources, their effects on human health, and control methods. Effects of heavy metals on plants, animals, and aquatic habitats are also discussed as these are consumed by human beings as food and are sources of heavy metals in the human body.

SOURCES OF HEAVY METALS

The sources of heavy metals are diverse in nature [31]. Many human activities such as smelting, use of insecticides, and burning of fossil fuel lead to severe environmental pollution. Many gases are emitted into the atmosphere as a result of the oxidation of metals present in the fuels. In aquatic systems, the presence of heavy metals is attributed to metal plating, chemical synthesis, catalysis, battery, fertilizer, paint, paper, and mining industries.

Industrial, agricultural, and domestic sources of heavy metal cause various health problems to habitats. In the mining area, activities such as smelting and mining have increased heavy metal pollution. Vehicle emission is a major source of arsenic, cadmium, cobalt, nickel, lead, antimony, and vanadium in air. Lead is used for pot glazing. Also, it is present in vessels and batteries. Due to the presence of lead in all these materials, lead enters the body through water resources, fruits, and vegetables. The use of lead in bullets, smoking, plumbing, and painting makes it a widely spread pollutant. The population near the industrial area is affected due to the disposal of heavy metal waste of nearby industries [31]. Rapid urbanization and tourism are the needs of the modern generation. But these touristic sites are prone to heavy metal pollution. The places where the cultivation of vegetable takes place are also heavy metal prone areas. Many activities related to cash crops and market gardening have increased the income of the developing population [32]. On the other hand, these activities have increased heavy metal presence in the crops and hence in the living organisms and human beings. Many pesticides used to increase crop production contains heavy metals [32]. Nowadays, discarded electronic, electrical, computer devices, and their components contribute to the solid waste in industrial and urban areas. This electronic waste (E-waste) contains many harmful heavy metals. Burning and incinerations of E-waste is one of the ways to treat these wastes [33].

The modern-day requirement of energy has encouraged the use of non-conventional and modified methods of energy generation. Nuclear fission is one of the important alternatives. Various radioactive isotopes are being handled in large quantities [34]. Also, soil pollution affects adversely not only human health but also the economic health of developing country like India. The level of heavy metals in soil is controlled by anthropogenic activities [35]. Heavy metal pollution affects air quality. The presence of heavy metal in the particulate matter makes them more harmful [36]. Shipyard and related industries use Aluminium. Also, tannery industries and other sources such as heavy traffic, fabrication of various equipment, various regular activities in hardware shops and workshops are major sources of heavy metal in aerosol in urban areas. These heavy metals enter through vegetables and accumulates in the human body [37 - 39].

Mercury is a specific heavy metal with some unique uses. The use of mercury in pesticides and deposition of industrial vent gases in the atmosphere causes mercury pollution. The use of CFL bulbs saves a lot of energy. On the other hand, it increases mercury pollution [40].

EFFECT ON HUMAN HEALTH

The presence of heavy metal in the air is harmful as they can enter the body system through air inhalation and cause respiratory diseases. Many investigations are reported on the effects of heavy metals on human beings [38 - 48]. Heavy metal consumption by fishes and other aquatic animals affects aquatic life and also people consuming them as food [49 - 54].

Human body functioning is severely affected by the presence of excess heavy metals. Both acute and chronic effects on nervous systems, liver, and kidney are observed due to the accumulation of heavy metals in the body [43, 55, 56]. Sometimes, they severely affect metabolic activities. Some metals like Aluminium can be removed by elimination activities. Due to this, they cause only short term threat. Free radical formation of heavy metals induces excess stress on the body system. Heavy metals displace originally bound heavy metal from protein structure; this causes malfunction of cells. Arsenic affects the function of cell respiration, enzyme activities, and mitosis. The presence of lead affects many critical biological activities [43, 56]. Being exposed to lead may result in the change of testicular functions in human beings [49 - 51].

For cadmium, smoking is the largest source of cadmium. Non-smokers and small children are affected by exposure. In industrial areas, where cadmium is present in considerable quantities, it may enter the human body through plants, milk, and seafood [34, 41, 42]. Cadmium has a tendency to accumulate and enter the food chain. The presence of cadmium in human body systems affect the functioning of the brain, kidney, liver, and nervous system [50, 51].

Mercury is present in many fishes and lean species. It has a tendency to accumulate in aquatic animals [54 - 56]. Old fatty fishes like shark, halibut, and redfish contain a considerable amount of mercury. It enters the human body through food, causing cardiovascular, haematological, pulmonary, renal, immunological, neurological, endocrine, reproductive, and embryonic toxicological health effects [57 - 59]. Other heavy metals like Nickel, Tin, and Manganese affects the nervous system [60, 61]. The health hazards caused by heavy metals are of two types. One is due to exposure to the heavy metal concentration above threshold limits. The other is due to exposure for a very long time to any concentration of heavy metals. These metals have very wide industrial applications. So exposure to these metals in the modern era of rapid economical and industrial development is unavoidable. Some metals like chromium, beryllium, nickel, cadmium, and arsenic are classified as carcinogenic [62]. The exact mechanism of carcinogenesis is not known as the nature of the interaction of these metals with biological systems is highly complex. The essential metals also can be carcinogenic with certain other elements and complexes. Target hazard quotients (THQ) are indicators of health risks associated with metals [63]. The presence of heavy metals contribute to high THQ values.

Advances in technology have helped us to get a better idea about the effects of heavy metal exposure. In Asian countries, the problem of heavy metal content in drinking water is a cause of concern [64, 65]. The population living near the zinc smelter is exposed to heavy metal pollution through multiple ways. Local farmers are largely exposed to heavy metal risk due to lead and cadmium in grains and vegetables *via* plant intake [66, 67]. Organic and inorganic forms of metal like lead are present in the atmosphere. Organic lead can be easily absorbed through the skin. Young children and pregnant women are susceptible to lead toxicity [68]. Many studies for heavy metal toxicity in fish indicated that the heavy metal concentration in fish is not above permissible limits near rivers in Iran [69]. Similar studies are reported by investigators from various countries. Studies show that haematological parameters in the freshwater fish are affected by heavy metals [70]. Consumption of the fish exposed to heavy metal is always harmful. Urinary excretion of cadmium and renal and bone biomarkers are associated [71]. They can be used for assessing the cadmium burden on the body [71]. In the human body, the highest concentration of cadmium is generally found in kidneys and livers. The consumption of staple food leads to the accumulation of heavy metals (cadmium) in the human body [71]. In most countries, 5 to 25 micrograms per day cadmium consumption is observed [71]. If we consider cigarette smoking, one cigarette contains 1-3 micrograms of cadmium. Smoking of one pack of cigarettes is comparable to intake from food [71]. In the case of inhalation, ten to fifty percent cadmium is absorbed. Cadmium absorption through the skin is negligible. In the kidney, cadmium shows adverse and detectable effects when the concentration exceeds a certain threshold value. 150-200 ppm is the threshold concentration at which renal effects are likely to occur [72 - 76]. Itai disease in Japan was a wakeup call regarding the effect of cadmium on bones. Studies in many other countries also have confirmed the bone effects of cadmium [72 - 76]. Alteration of genes related to stem and somatic cells can be induced by metals like aluminium, arsenic, chromium, nickel, and selenium [77]. Regular antioxidant use can reduce the toxic effects of these metals. Modification of soil in urban and suburban areas because of human activities exposes the increased population to heavy metal pollution [78].

REMOVAL METHODS

Many treatment technologies for wastewater for the removal of heavy metal are available. Many efforts are focused on finding an easier and more economical method for heavy metal removal from wastewater. Methods like adsorption, ion exchange, electrodialysis, and coagulation use fundamental principles of physics and chemistry. The selection of removal methods depends upon the influent concentration, acceptable limit, and amount of effluent. Biological treatment methods like advanced oxidation, trickling filters, aeration, activated sludge, aerobic, and anaerobic digestion can be used for heavy metal treatment. In biological processes, microorganisms play a major role in settling the impurities [69]. Toxic inorganic impurities are normally removed by chemical treatment methods. Adsorption by various materials is the most popular research method [79, 80]. According to Gautam *et al.*, remedial technologies for heavy metal removal can be classified as precipitation and coagulation, ion exchange, membrane filtration, bioremediation, heterogeneous photo-catalysts, and adsorption [81].

Many investigators have investigated the removal of heavy metal by using adsorbents derived from waste materials like biomass, leaves, and fruit waste [82]. Many biological materials isolated from wood waste and leaves are investigated for heavy metal removal [83]. Agricultural waste material has been used in many investigations as raw material for the removal of heavy metals from wastewater [84, 85]. Many investigations on adsorption for heavy metals are focused on the parameters like pH, initial concentration, adsorbent dose, and particle size of adsorbent [86 - 93]. Packed bed experiments have been carried out to study the practical applicability of adsorption [94]. These studies are often aimed at studying the effect of flow rate, bed height, and initial concentration on the breakthrough curve. Greener and environment-friendly methods are gaining importance. From this aspect, adsorption by using agricultural waste is one of the most sought after method for heavy metal removal [95]. Bio-sorption using commonly observed species like bacteria and yeast has been also effective for heavy metal removal from wastewater [96]. Many heavy metals can be removed effectively by using carbon nanotubes [97]. Alumino-silicate, chitosan, zeolite, chitin derivatives, natural polyelectrolyte, and many agricultural waste materials are being investigated for their ability to remove heavy metals [98 - 109].

Trickling filters with different types of support packings can be used for the removal of heavy metals [110]. Investigations by using aerobic and anaerobic pathways for heavy metal removal have yielded more than 90 percent removal [111 - 113]. Biological siliceous filter with active biomass can be used for heavy metal removal [114]. Many scientific and engineering principles such as bio-sorption, immobilization, membrane bioprocesses, and activated sludge method are explored for the removal of heavy metals from wastewater with varying amounts of success [115 - 129].

Filtration methods such as ultrafiltration and nano-filtration can also be used [28, 92, 93]. Many new coagulant materials are being tried for the removal of heavy metals [129].

Gering and Scamehorn carried out electro-dialysis for heavy metal removal [130]. They studied the effect of parameters like current efficiency, stack resistance, and osmotic water transfer on heavy metal removal. They observed that with an increase in current density, the removal percentage increases. Mahvi and

Bazrafshan investigated the electro-coagulation process for the cadmium electrode [129]. They observed that the initial concentration of cadmium affects removal efficiency.

Separation of zinc and cadmium ions from sulphate solution by membrane transport and ion floatation was investigated by Ulewiz and Walkowiak [131]. They used ion floatation experiments for dilute aqueous solutions with an anionic surfactant (sodium dodecylbenzene sulfonate) and a cationic surfactant (hexadecylpyridinium chloride). They observed that there was an increase in floatation separation of cadmium and zinc with an increase in sulphate concentration Adsorbing particle floatation was used for the removal of cadmium from wastewater by Kobayashi [132]. He used bentonite and a catholic surfactant. The results indicated that the addition of polyacryl amide increased floatation efficiency.

Membrane chromatography was used for the removal of heavy metal from the aquatic solution by Denzili *et al.* [133]. Their studies suggested that the heavy metal could be repeatedly adsorbed and stripped without significant losses. Comparative studies were done by Qdais and Moussa on the removal of heavy metals by membrane technology [134]. They used nanofiltration and reverse osmosis for the removal of heavy metals from the wastewater. They observed that reverse osmosis was more effective than nanofiltration.

Solvent extraction is a suitable method for the removal of heavy metals from industrial effluent. Various organic solvents selectively extract heavy metals from solutions. It is possible to obtain more than 99 percent removal by using organic solvents [135, 136].

CONCLUSION

Many human activities such as smelting, use of insecticides, and burning of fossil fuel lead to severe environmental pollution. Many gases are emitted into the atmosphere as a result of the oxidation of metals present in the fuels. In aquatic systems, the presence of heavy metals is attributed to metal plating, chemical synthesis, catalysis, battery, fertilizer, paint, paper, and mining industries. Metal like cadmium can cause renal dysfunction and bone weakness [137 - 139]. Heavy metals are harmful for human beings as they adversely affect kidney and liver functioning. There are many other acute and chronic effects caused by heavy metals. Physical methods such as filtration and adsorption (physio-chemical) are simple and effective. These methods can be used to remove heavy metals from wastewater when they are present in very high concentration. Depending on the influent concentration, acceptable limit, and amount of effluent, the removal method can be selected. Chemical treatment methods like coagulation, chlorination, flocculation, and precipitation can be used in many applications due to their simplicity and robustness. Sorption methods are the most widely explored treatment techniques for the removal of heavy metals from water.

CONSENT FOR PUBLICATION

Not applicable.

CONFLICT OF INTEREST

The authors confirm that the contents of this chapter have no conflict of interest.

ACKNOWLEDGEMENT

The author is indebted to the management of the affiliating institute for encouragement.

REFERENCES

[1] Rolka E, Żołnowski AC, Sadowska MM. Assessment of heavy metal content in soils adjacent to the DK16-Route in olsztyn (North-Eastern Poland). Pol J Environ Stud 2020; 29(6): 4303-11.
[http://dx.doi.org/10.15244/pjoes/118384]

[2] Rattan RK, Datta SP, Chandra S. Heavy metals and environmental quality. Fert News 2002; 47(11): 21-40.

[3] Zwolak A, Sarzyńska M, Szpyrka E. Sources of soil pollution by heavy metals and their accumulation in vegetables: a review. Water Air Soil Pollut 2019; 230: 164.

[4] Wiseman CL, Zereini SF, Püttmann W. Metal and metalloid accumulation in cultivated urban soils: a medium-term study of trends in Toronto, Canada. Sci Total Environ 2015; 15: 564-72.

[5] Karimi R, Ayoubi S, Jalalian A, Sheikh-Hosseini AR, Afyuni M. Relationships between magnetic susceptibility and heavy metals in urban topsoils in the arid region of Isfahan, central Iran. J Appl Geophys 2011; 74(1): 1-7.
[http://dx.doi.org/10.1016/j.jappgeo.2011.02.009]

[6] Hanesch M, Scholger R. The influence of soil type on the magnetic susceptibility measured throughout soil profiles. Geophys J Int 2005; 161(1): 50-6.
[http://dx.doi.org/10.1111/j.1365-246X.2005.02577.x]

[7] Dankoub Z, Ayoubi S, Khademi H, Sheng-Gao LU. Spatial distribution of magnetic properties and selected heavy metals in calcareous soils as affected by land use in the Isfahan region. Central Iran, Pedosphere 2012; 22(1): 33-47.
[http://dx.doi.org/10.1016/S1002-0160(11)60189-6]

[8] Naimi S, Ayoubi S. Vertical and horizontal distribution of magnetic susceptibility and metal contents in an industrial district of central Iran. J Appl Geophys 2013; 96: 55-66.
[http://dx.doi.org/10.1016/j.jappgeo.2013.06.012]

[9] Taghipour M, Ayoubi S, Khademi H. Contribution of lithologic and anthropogenic factors to surface soil heavy metals in western Iran using multivariate geostatistical analyses. Soil Sediment Contam 2011; 20(8): 921-37.
[http://dx.doi.org/10.1080/15320383.2011.620045]

[10] Dearing JA, Livingstone IP, Bateman MD, White KK. Palaeoclimate records from OIS 8.0–5.4 recorded in loess-palaeosol sequences on the Matmata Plateau, southern Tunisia, based on mineral magnetism and new luminescence dating. Quat Int 2001; 76–77: 43-56.
[http://dx.doi.org/10.1016/S1040-6182(00)00088-4]

[11] Petrovsk'y E, Ellwood BB. Magnetic monitoring of air-, land-, and water-pollution. Quaternary Climates, Environments and Magnetism. Cambridge: University press 1999; pp. 279-322.
[http://dx.doi.org/10.1017/CBO9780511535635.009]

[12] Dankoub Z, Khademi H, Ayoubi S. Magnetic susceptibility and its relationship with the concentration of selected heavy metals and soil properties in surface soils of the Isfahan region. J Environ Stud (Northborough) 2012; 38(63): 17-26.

[13] Li C, Zhou K, Qin W, *et al.* A review on heavy metals contamination in soil: effects, sources, and remediation techniques. Soil Sediment Contam 2019.

[14] Kanu MO, Meludu OC, Oniku SA. Comparative study of top soil magnetic susceptibility variation based on some human activities. Geofis Int 2014; 53(4): 411-23.
[http://dx.doi.org/10.1016/S0016-7169(14)70075-3]

[15] Ayoubi S, Jabbari M, Khademi H. Multiple linear modeling between soil properties, magnetic susceptibility and heavy metals in various land uses. Model Earth Syst Environ 2018; 4(2): 579-89.
[http://dx.doi.org/10.1007/s40808-018-0442-0]

[16] Ayoubi S, Soltani Z, Khadem H. Particle size distribution of heavy metals and magnetic susceptibility in an industrial site. Bull Environ Contam Toxicol 2018; 100(5): 708-14.
[http://dx.doi.org/10.1007/s00128-018-2316-6]

[17] Ayoubi S, Adman V, Yousefifard M. Use of magnetic susceptibility to assess metals concentration in soils developed on a range of parent materials. Ecotoxicol Environ Saf 2019; 168: 138-45.
[http://dx.doi.org/10.1016/j.ecoenv.2018.10.024]

[18] Ayoubi S, Karami M. Pedotransfer functions for predicting heavy metals in natural soils using magnetic measures and soil properties. J Geochem Explor 2019; 197: 212-9.
[http://dx.doi.org/10.1016/j.gexplo.2018.12.006]

[19] Afshari A, Khademi H, Ayoubi S. Risk assessment of heavy metals contamination in soils and selected crops in zanjan urban and industrial regions. J Water Soil 2015; 29(1): 151-63. [Agricultural Sciences and Technology].

[20] Naeimi M, Ayoubi S, Azimzadeh B. Use of multivariate statistics and geostatistics to differentiate the lithologic and anthropogenic sources of some heavy metals in Zobahan Industrial District, Isfahan Province. J Water Soil Agric Sci Technol 2013; 27(3): 560-9.

[21] Musa B, Abdullahi MS. The toxicological effects of cadmium and some other heavy metals in plants and humans. J Environ Sci Water Res 2013; 2(8): 245-9.

[22] Singh JA, Kalamdhad A. Effects of heavy metals on soil, plants, human health and aquatic life. Int J Res Chem Environ 2011; 1(2): 15-21.

[23] Khayatzadeh J, Abbasi E. The Effects of Heavy Metals on Aquatic Animals. The 1st International Applied Geological Congress, Department of Geology, Islamic Azad University - Mashad Branch, Iran, 26-28, April 2010, pp.688-695.

[24] Bada BS, Oyegbami OT. Heavy metals concentrations in roadside dust of different traffic density. J Environ Earth Sci 2012; 2(8): 54-61.

[25] Kumar U. Agricultural products and by-products as a low cost adsorbent for heavy metal removal from water and wastewater: A review. Sci Res Essay 2006; 1: 33-7.

[26] Palit S. Wastewater treatment and desalination with the help of membrane separation processes-a short and far-reaching review. Int J Chem Sci Appl 2012, 3(3): 352-355.

[27]   Nogueira CA, Delmas F. New flow sheet for the recovery of cadmium, cobalt and nickel from spent Ni–Cd batteries by solvent extraction. Hydrometallurgy 1999; 52: 267-87.
[http://dx.doi.org/10.1016/S0304-386X(99)00026-2]

[28]   Fu F, Wang Q. Removal of heavy metal ions from wastewaters: a review. J Environ Manage 2011, 92(3): 407-418.

[29]   Kulkarni SJ, Kaware JP. Removal of cadmium from wastewater by groundnut shell adsorbent-batch and column studies. Int J Chem Eng Res 2014; 6(1): 27-37.

[30]   Hydari S, Sharififard H, Nabavinia M, Parvizi M. A comparative investigation on removal performances of commercial activated carbon, chitosan biosorbent and chitosan/activated carbon composite for cadmium. Chem Eng J 193: 276-82.
[http://dx.doi.org/10.1016/j.cej.2012.04.057]

[31]   Rajeswari TR, Sailaja N. Impact of heavy metals on environmental pollution, National Seminar on Impact of Toxic Metals, Minerals and Solvents leading to Environmental Pollution. J Chem Pharm Sci 2014; 175-81.

[32]   Hounkpe JB, Kelome NC, Adèchina R, Lawani RN. Assessment of heavy metals contamination in sediments at the lake of Aheme in southern of Benin (West Africa). J Mater Environ Sci 2017; 8(12): 4369-77.
[http://dx.doi.org/10.26872/jmes.2017.8.12.460]

[33]   Fosu-Mensah BY, Addae E, Yirenya-Tawiah D, Nyame F. Heavy metals concentration and distribution in soils and vegetation at Korle Lagoon area in Accra, Ghana Cogent Environ 2017; 3(1)
[http://dx.doi.org/10.1080/23311843.2017.1405887]

[34]   Naja GM, Volesky B. Toxicity and Sources of Pb, Cd, Hg, Cr, As, and radionuclides in the Environment.Heavy Metals in the Environment. Ottawa, ON, Canada: Environ. Canada 2009; pp. 1-50.

[35]   Ladachart R, Sutthirat C, Hisada K, Charusiri P. Soil Erosion and Heavy Metal Contamination in the Middle Part of the Songkhla Lake Coastal Area, Southern Thailand.Coastal Environments:Focus on Asian Regions. Dordrecht: Springer 2012.
[http://dx.doi.org/10.1007/978-90-481-3002-3_8]

[36]   Aksu A. Sources of metal pollution in the urban atmosphere (A case study: Tuzla, Istanbul). J Environ Health Sci Eng 2015; 13: 79.
[http://dx.doi.org/10.1186/s40201-015-0224-9]

[37]   Alam M G M, Snow E T, Tanaka A. Arsenic and heavy metal contamination of vegetables grown in Santa village, Bangladesh. Sci Total Environ 2003; 308: 83-96.

[38]   Singh R, Ahirwar NK, Tiwari J, Pathak J. Review on sources and effect of heavy metal in soil: its bioremediation. Int J Res Appl Nat Soc Sci 2018; 1-22.

[39]   Chouhan B, Meena P, Poonar N. Effect of heavy metal ions in water on human health. Int J Sci Eng Res 2016; 4(12): 32p.

[40]   Xu X, Zhao Y, Zhao X, Deng WJ, Wang Y. Sources of heavy metal pollution in agricultural soils of a rapidly industrializing area in the Yangtze Delta of China. Ecotoxicol Environ Saf 2014; 108: 161-7.
[http://dx.doi.org/10.1016/j.ecoenv.2014.07.001]

[41]   Laniyan TA, Kehinde Phillips OO, Elesha L. Hazards of heavy metal contamination on the groundwater around a municipal dumpsite in Lagos, Southwestern Nigeria. IACSIT Int J Eng Technol 2011; 11(5): 61-70.

[42]   Martin S. Human health effects of heavy metals. Environ Sci Tech Briefs for Citizens 2009; 15: 1-6.

[43]   Jaishankar M, Tseten T, Anbalagan N, Mathew BB, Beeregowda KN. Toxicity, mechanism and health effects of some heavy metals. Interdiscip Toxicol 2014; 7(2): 60-72.
[http://dx.doi.org/10.2478/intox-2014-0009]

[44]   Pizent A, Tariba B, Zivkovic T. Reproductive toxicity of metals in men. Arh Hig Rada Toksikol 2012; 63 (Suppl. 1): 3546.
[http://dx.doi.org/10.2478/10004-1254-63-2012-2151]

[45]   Dee KH, Abdullah F, Md Nasir S, Appalasamy S, Ghazi RM, Rak AE. Health risk assessment of heavy metals from smoked *Corbicula fluminea* collected on roadside vendors at kelantan, Malaysia. BioMed Res Int 2019; 20199596810
[http://dx.doi.org/10.1155/2019/9596810]

[46]   Rothman JA, Leger L, Kirkwood J S, McFrederick QS. Cadmium and selenate exposure affects the honey bee microbiome and metabolome, and bee-associated bacteria show potential for bioaccumulation. Appl Environ Microbiol 2019; 85(21): 1411-1419.
[http://dx.doi.org/10.1128/AEM.01411-19]

[47]   Sow AY, Ismail A, Zulkifli SZ, Amal MN, Hambali KA. Survey on heavy metals contamination and health risk assessment in commercially valuable asian swamp Eel, *Monopterus albus* from Kelantan, Malaysia. Sci Rep 2019; 9: 6391.
[http://dx.doi.org/10.1038/s41598-019-42753-2]

[48]   McNeill RV, Mason AS, Hodson ME, Catto JWF, Southgate J. Specificity of the Metallothionein-1 Response by Cadmium-Exposed Normal Human Urothelial Cells. Int J Mol Sci 2019; 20(6): 1344.
[http://dx.doi.org/10.3390/ijms20061344]

[49]   Sridhar N, Senthilkumar JS, Subburayan MR. Removal of toxic metals (lead & copper) from automotive industry waste water by using fruit peels. Int J Adv Inform Commun Tech 2014; 1(2): 187-92.

[50]   Chibbar S, Sharma N. A review on impact of heavy metal toxicity on environment. Int J Innov Res Adv Stud 2014; 3(5): 530-41.

[51]   Khan S, Cao Q, Zheng Y M, Huang Y Z, Zhu Y G. Health risks of heavy metals in contaminated soils and food crops irrigated with wastewater in Beijing, China. Environ Pollut 2008; 152:686-692.

[52]   Pandey G, Madhuri S. Heavy metals causing toxicity in animals and fishes. Res J Anim Veter Fishery Sci 2014; 2(2): 17-23.

[53]   Agbozu IE, Ekweozor IKE, Opuene K. Survey of heavy metals in the catfish Synodontis clarias. J Environ Sci Toxicol Food Technol 2007; 4(1): 93-7.
[http://dx.doi.org/10.1007/BF03325966]

[54]   Isienyi NC, Ipeaiyeda AR. Heavy metal effect on well water and biodiversity near Lapite Dumpsite in Ibadan, Oyo State, Nigeria. J Environ Sci Toxic Food Tech 2014; 8(2): 27-30.
[http://dx.doi.org/10.9790/2402-08232730]

[55]   Jung MC. Heavy metal concentrations in soils and factors affecting metal uptake by plants in the vicinity of a Korean Cu-W Mine. Sensors (Basel) 2008; 8: 2413-23.
[http://dx.doi.org/10.3390/s8042413]

[56]   Ziemacki G, Viviano G, Merli F. Heavy metals: sources and environmental presence. Ann. 1st Super Sanita 1989; 25(3): 531-6.

[57]   Cheng S. Effects of heavy metals on plants and resistance mechanisms. Environ Sci Pollut Res Int 2003; 10(4): 256-64.
[http://dx.doi.org/10.1065/espr2002.11.141.2]

[58]   Rice Kevin, Walker Ernest, Wu Miaozong, Gillette Christopher, Blough Eric. Environmental mercury and its toxic effects. J Prev Med Public Health 2014; 47: 74-83.
[http://dx.doi.org/10.3961/jpmph.2014.47.2.74]

[59]   Morais S. F.G.e-Costa, "Heavy Metals and Human Health", . https:// www.researchgate. net /publication /221923928

[60]   Brochin R, Leone S, Phillips D, Shepard N, *et al.* The cellular effect of l ead poisoning and its clinical

picture. Georgetown Univ J Health Sci 5(2): 1-8.

[61]   Mahurpawar M. Effects of heavy metals on human health. Int J Res Soc Issues Environ Prob 2015; 1-7.

[62]   Goyer R, Golub M, Choudhury H, Hughes M, Kenyon E, Stifelman M. Issue paper on the human health effects of metals US Environmental Protection Agency Risk Assessment Forum 1200 Pennsylvania Avenue, NW. 2004; p. 49.

[63]   Mudgal V, Madaan N, Mudgal A, Singh RB, Mishra S. Effect of toxic metals on human health. Open Nutraceuticals J 2010; 3: 94-9.
       [http://dx.doi.org/10.2174/18763960010030100094]

[64]   Kulkarni SJ, Dhokpande SR, Kaware JP. A Review on studies on effect of heavy metals on man and environment. Int J Res Appl Sci Eng Technol 2014; 2(10): 227-9.

[65]   Singh A, Sharma RK, Agrawal M, Marshall FM. Risk assessment of heavy metal toxicity through contaminated vegetables from waste water irrigated area of Varanasi, India. Trop Ecol 2010; 51(2S): 375-87.

[66]   Guerra F, Trevizam AR, Muraoka T, Marcante NC, Canniatti-Brazaca SG. Heavy metals in vegetables and potential risk for human health. Sci Agric 2012; 69(1): 54-60.
       [http://dx.doi.org/10.1590/S0103-90162012000100008]

[67]   Baghaie AH, Fereydoni M. The potential risk of heavy metals on human health due to the daily consumption of vegetables. Environ Health Eng Manag 2019; 6(1): 11-6.
       [http://dx.doi.org/10.15171/EHEM.2019.02]

[68]   Indian National Science Academy. Hazardous metals and minerals pollution in India A Position Paper 2011; 2-2.

[69]   Mortazavi S, Fard PN. Risk assessment of non-carcinogenic effects of heavy metals from Dez river fish. Iranian J Health Sci 2017; 5(4): 10-25.
       [http://dx.doi.org/10.29252/jhs.5.4.10]

[70]   Vinodhini R, Narayanan M. The impact of toxic heavy metals on the haematological parameters in common carp (Cyprinus Carpio l.). Iran J Environ Health Sci Eng 2009; 6(1): 23-8.

[71]   Bernard A. Cadmium & its adverse effects on human health. Indian J Med Res 2008; 128: 557-64.

[72]   Sankhla MS, Kumari M, Nandan M, v Kumar R, Agrawal P. Heavy metals contamination in water and their hazardous effect on human health-a review. Int J Curr Microbiol Appl Sci 2016; 5(10): 759-66.
       [http://dx.doi.org/10.20546/ijcmas.2016.510.082]

[73]   Duruibe JO, Ogwuegbu MO, Egwurugwu JN. Heavy metal pollution and human biotoxic effects. Int J Phys Sci 2007; 2(5): 112-8.

[74]   Malik QA, Khan MS. Effect on human health due to drinking water contaminated with heavy metals. J Pollut Eff Cont 2016; 5(1)

[75]   Shen X, Chi Y, Xiong K. The effect of heavy metal contamination on human and animal health in the vicinity of a zinc smelter Plos One 2019; 14(10): 1-15.
       [http://dx.doi.org/10.1101/459644]

[76]   MahMoud MAM, Abdel-Mohsein HS. Health risk assessment of heavy metals for Egyptian population *via* consumption of poultry edibles. Adv Anim Vet Sci 2015; 3(1): 58.
       [http://dx.doi.org/10.14737/journal.aavs/2015/3.1.58.70]

[77]   Mishra S, Dwivedi SP, Singh RB. A Review on epigenetic effect of heavy metal carcinogens on human health. Open Nutraceut J 2010; 3: 188-93.
       [http://dx.doi.org/10.2174/18763960010030100188]

[78]   Giuffre L, Romaniuk RI, Marban L, Rois RP, Garcai Torres TP. Public health and heavy metals in urban and periurban horticulture. Emir J Food Agric 2012; 24(2): 148-54.

[79]   Gunatilake SK. Methods of removing heavy metals from industrial wastewater. J Multidiscip Eng Sci Studi 2015; 1(1): 12-18.

[80]   Renu B, Agarwal M, Singh K. Methodologies for removal of heavy metal ions from wastewater: an overview. Interdiscip Environ Rev 2017; 18(2): 124-43.
[http://dx.doi.org/10.1504/IER.2017.087915]

[81]   Gautam RK, Sharma SK, Mahiya S, Chattopadhyaya MC. Contamination of heavy metals in aquatic media:transport, toxicity and technologies for remediation from book Heavy Metals In Water: Presence, Removal and Safety 2014.http://pubs.rsc.org

[82]   Abdel-Raouf MS, Abdul-Raheim ARM. Removal of heavy metals from industrial waste water by biomass- based materials: A review. J Pollut Eff Cont 2017; 5: 1.
[http://dx.doi.org/10.4172/2375-4397.1000180]

[83]   Salehzadeh J. Removal of heavy metals Pb^{2+}, Cu^{2+}, Zn^{2+}, Cd^{2+}, Ni^{2+}, Co^{2+} and Fe^{3+} from aqueous solutions by using xanthium pensylvanicum. Leonardo J Sci 2013; (23): 97-104.

[84]   Lakherwal D. Adsorption of heavy metals: a review. Int J Environ Res Dev 2014; 4(1): 41-8.

[85]   Shaikh RB, Saifullah B, Rehman F. Greener method for the removal of toxic metal ions from the wastewater by application of agricultural waste as an adsorbent. Water 2018; 10(1316): 14p.
[http://dx.doi.org/10.3390/w10101316]

[86]   Muthulakshmi AN, Anuradha J. Removal of cadmium ions from water/waste water using chitosan - a review, research & reviews. Int J Eco Envirom Sci 2015; S1: 9-15.

[87]   Kulkarni S J, Kaware J P. Cadmium removal by adsorbent prepared from local agricultural waste of rice processing. Int J Sci Res Chem Eng 2015; 2(1): 014-22.

[88]   Mwanyika FT, Ogendi GM, Kipkemboi JK. Removal of heavy metals from wastewater by a constructed wetland system at Egerton University, Kenya. J Environ Sci Toxico Food Tech 2016; 10(1): 15-20.

[89]   Venkatesan G, Senthilnathan U. Adsorption batch studies on the removal of cadmium using wood of derris indica based activated carbon. Res J Chem Environ 2013; 17(5): 19-25.

[90]   Singh N, Gupta SK. Adsorption of heavy metals: A review. Int J Innov Res Sci Eng Tech 2016; 5(2): 2267-88.

[91]   Tadepalli KSR. Murthy, "Comparison studies for copper and cadmium removal from industrial effluents and synthetic solutions using mixed adsorbent in batch mode. Int J Chemtech Res 2017; 10(5): 652-63.

[92]   Kulkarni SJ. A review on recent (2013 -2017) investigations for cadmium removal from wastewater. Galore Int J Appl Sci Hum 2017; 1(3): 1-6.

[93]   Kulkarni SJ. Effect of presence of other heavy metals on removal of cadmium by low cost adsorbents. Int J Latest Eng and Manage Res 2017; 2(9): 15-20.

[94]   Radaideh J A, Abdulgader H A, Barjenbruch M. Evaluation of absorption process for heavy metals removal found in pharmaceutical wastewater. J Med Toxico Clin Forens Med 2017; 3(2:9): 1698-9465.
[http://dx.doi.org/10.21767/2471-9641.100029]

[95]   Kulkarni SJ, Kaware JP. Packed bed adsorption column modeling for cadmium removal. Int J Therm Environ Eng 2015; 9(2): 75-82.

[96]   Sharma S, Rana S, Thakkar A, Baldi A, Murthy RSR. Physical, chemical and phytoremediation technique for removal of heavy metals. J Heavy Metal Toxic Dis 2016; 1(2:10): 1-15.

[97]   Ruparelia JP, Duttagupta SP, Chatterjee AK, Mukherji S. Potential of carbon nanomaterials for removal   of   heavy   metals   from   water   http://dspace.library.iitb.ac.in /jspui/bitstream/10054/1376/1/5238.pdf

[http://dx.doi.org/10.1016/j.desal.2007.08.023]

[98] Tariq W, Saifullah M, Anjum T, Javed M, Tayyab N, Shoukat I. Removal of heavy metals from chemical industrial wastewater using agro based bio-sorbents. Acta Chemica Malaysia 2018; 2(2): 9-14. [ACMY].

[99] Ayres DM, Davis A P, Gietka P M. Removing heavy metals from wastewater University of Maryland, Engineering Research Center Report 21.

[100] Joshi NC. Heavy metals, conventional methods for heavy metal removal, biosorption and the development of low cost adsorbent. Eur J Pharm Med Res 2017; 4(2): 388-93.

[101] Zhang P, Ding W, Zhang Y, Dai K, Liu W. Heavy metal ions removal from water using modified zeolite. J Chem Pharm Res 2014; 6(11): 507-14.

[102] Wambu EW, Attahiru S, Shiundu PM, Wabomba J. Removal of heavy-metals from wastewater using a hydrous alumino-silicate mineral from Kenya. Bull Chem Soc Ethiop 2018; 32(1): 39-51.
[http://dx.doi.org/10.4314/bcse.v32i1.4]

[103] Jain G. Removal of copper and zinc from wastewater using chitosan Department of Chemical Engineering, MSc thesis, National Institute of Technology, Rourkela 2013; 58.

[104] Acharya J, Kumar U, Rafi PM. Removal of heavy metal ions from wastewater by chemically modified agricultural waste material as potential adsorbent-a review. Int J Curr Sci Eng Tech 2018; 8(3): 526-30.
[http://dx.doi.org/10.14741/ijcet/v.8.3.6]

[105] Peters RW, Shem L. Separation of heavy metals: removal from industrial wastewaters and contaminated soil 1993.https://www.osti.gov/servlets/purl/6504209/

[106] Mohameda SH. A.d A. El-Gendya, A.l H. Abdel-kadera, and E. A. El-Ashkarc, "Removal of heavy metals from water by adsorption on chitin derivatives. Pharma Chem 2015; 7(10): 275-83.

[107] Verma A, Rejendra K, Singh K, Shukla S. Use of low cost adsorbents for the remediation of heavy metals from waste water. Int J Latest Tech Eng Manage Appl Sci 2017; 6(7): 13-20.

[108] Mane PC, Bhosle AB, Jangam CM, Mukate SV. Heavy metal removal from aqueous solution by Opuntia: A Natural Polyelectrolyte J Nat Prod Plant Resour 1(1): 75-80.

[109] Dukare GR, Bhoir A, Raut S, Parkar P. Removal of heavy metals by means of banana and orange peels. Int J Innov Res Sci Eng Technol 2018; 7(4): 3466-73.

[110] Kulkarni SJ, Goswami AK. Isotherm, kinetics and trickling flow studies for removal of chromium from synthetic effluent by using mixed fruit peels (MFP). Int J Innov Tech Ex Eng 2019; 8(6S): 430-432.

[111] Mockaitis G, Rodrigues JA, Foresti E, Zaiat M. Toxic effects of cadmium (Cd^{2+}) on anaerobic biomass: kinetic and metabolic implications. J Environ Manage 2012; 106: 75-84.
[http://dx.doi.org/10.1016/j.jenvman.2012.03.056]

[112] Lopez-Perez PA, Neria-Gonzalez MI, Flores-Cotera LB. A mathematical model for cadmium removal using a sulfate reducing bacterium: Desulfovibrio alaskensis 6SR. Int J Environ Res 2013; 7: 501-12.

[113] Wang CL, Maratukulam PD, Lum AM, Clark DS, Keasling JD. Metabolic engineering of an aerobic sulfate reduction pathway and its application to precipitation of cadmium on the cell surface. Appl Environ Microbiol 2000; 66: 497-4502.
[http://dx.doi.org/10.1128/AEM.66.10.4497-4502.2000]

[114] R.A Deyanati tilki, M Shariat, "Study on removal of cadmium from water by bacterial biomass in biological siliceous filter. Majallah-i Danishgah-i Ulum-i Pizishki-i Mazandaran 2003; 13: 17-26.

[115] Hong AH, Burmamu BR, Umaru AB, Sadiq UM. Removal of heavy metals from contaminated water using Moringa Olefeira seed coagulant in Yola and its environs. Int J Eng Sci Invention 2017; 6(11): 40-5.

[116] Dhokpande S, Kaware J, Kulkarni S. Activated sludge for heavy metal removal- a review. Int J Res Appl Sci Eng Technol 2014; 2(7): 254-9.

[117] Maximous NN, Nakhla GF, Wan WK. Removal of heavy metals from wastewater by adsorption and membrane processes: a comparative study. Int J Scho Sci Res Innov 2010; 4(4): 125-30.

[118] Abdel-Aziz MH, Bassyouni M, Soliman MF, Gutub SA, Magram SF. Removal of heavy metals from wastewater using thermally treated sewage sludge adsorbent without chemical activation. J Mat Environ Sci 2017; 8(5): 1737-47.

[119] Ali SM, Galal A, Atta NF, Shammakh Y. Toxic heavy metal ions removal from wastewater by Nano Magnetite: case study Nile river water. Egypt J Chem 2017; 60(4): 601-12.
[http://dx.doi.org/10.21608/ejchem.2017.3583]

[120] Renu B, Agarwal M, Singh K. Heavy metal removal from wastewater using various adsorbents. J Water Reuse Desalin 2017; 388-420.https://iwaponline.com/jwrd/article-pdf/7/4/387/375972 /jwrd 00703 87.pdf
[http://dx.doi.org/10.2166/wrd.2016.104]

[121] Shafiq M, Alazba AA, Amin MT. Removal of heavy metals from wastewater using date palm as a biosorbent: a comparative review. Sains Malays 2018; 47(1): 35-49.
[http://dx.doi.org/10.17576/jsm-2018-4701-05]

[122] Muzenda E, Kabuba J, Ntuli F, Mollagee M, Mulaba Ae F. Cu (II) removal from synthetic waste water by ion exchange process. Proceedings of the World Congress on Engineering and Computer Science 2011; II: 5.

[123] Igwe JC. A review of potentially low cost sorbents for heavy metal removal and recovery. Terrestrial Aqua Environ Toxico 2007; 1(2): 60-9.

[124] Gangadhar G, Maheshwari U. Application of nanomaterials for the removal of pollutants from effluent streams. Nanosci Nanotechnol Asia 2012; 2: 140-50.
[http://dx.doi.org/10.2174/2210681211202020140]

[125] Al-Qahtani KM. Water purification using different waste fruit cortexes for the removal of heavy metals. J Taibah Univ Sci 2016; 10(5): 700-8.
[http://dx.doi.org/10.1016/j.jtusci.2015.09.001]

[126] Nacke H, Gonçalves AC Jr, Campagnolo MA, Coelho GF, Schwantes D, dos Santos MG. D. Luiz Briesch Jr., J.Zimmermann, "Adsorption of Cu (II) and Zn (II) from water by Jatropha curcas L. as biosorbent. Open Chem 2016; 14: 103-17.

[127] Mirbagheri SA, Biglarijoo N, Keyhannejad M. Pilot plant studies for the removal of heavy metals from industrial wastewater using adsorbents. Turkish J Eng Env Sci 2014; 38: 159-66.
[http://dx.doi.org/10.3906/muh-1404-18]

[128] Phadtare MJ, Patil ST. Removal of heavy metal from industrial waste water. Int J Adv Engg Res Studies 2015; 4(3): 4-8.

[129] Mahvi AH, Bazrafshan E. Removal of cadmium from industrial effluents by electro coagulation process using aluminum electrodes. World Appl Sci J 2007; 2: 34-9.

[130] Gering KL, Scamehorn JF. Use of electro dialysis to remove heavy metals from water. Sep Sci Technol 1988; 23: 2231-67.
[http://dx.doi.org/10.1080/01496398808058452]

[131] Ulewicz M, Walkowiak W. Separation of zinc and cadmium ions from sulfate solutions by ion flotation and transport through liquid membranes. Physicochem Probl Miner Proces 2003; 37: 77-86.

[132] Kobayashi K. Studies of the removal of Cd^+ ions by adsorbing particle floatation. Bull Chem Soc Jpn 1975; 48: 1750-4.
[http://dx.doi.org/10.1246/bcsj.48.1750]

[133] Denizli A, Say R, Arica Y. Removal of heavy metal ions from aquatic solutions by membrane chromatography. Separ Purif Tech 2000; 21: 181-90.
[http://dx.doi.org/10.1016/S1383-5866(00)00203-3]

[134] Abu Qdais H, Moussa H. Removal of heavy metals from wastewater by membrane processes: a comparative study. Desalination 2004; 164: 105-10.
[http://dx.doi.org/10.1016/S0011-9164(04)00169-9]

[135] Kulkarni SJ, Kaware JP. A review on research for cadmium removal from effluent. Int J Eng Sci Innov Technol 2013; 2(4): 465-9.

[136] Nogueira CA, Delmas F. New flow sheet for the recovery of cadmium, cobalt and nickel from spent Ni–Cd batteries by solvent extraction. Hydromettalurgy 1999; 52: 267-87.
[http://dx.doi.org/10.1016/S0304-386X(99)00026-2]

[137] Nordberg G, Jin T, *et al.* Low bone density and renal dysfunction following environmental cadmium exposure in China. Ambio. 2002; 31: p. 478-81.

[138] Akesson A, Bjellerup P, Lndh T, *et al.* Cadmium-induced effects on bone in population-based study of women. Environ Health Perspect 2006; 114: 830-834.
[http://dx.doi.org/10.1289/ehp.8763]

[139] Jarup L, Alfven T. Low level cadmium exposure, renal and bone effects-the OSCAR study. Biometals 2004; 17: 505-9.
[http://dx.doi.org/10.1023/B:BIOM.0000045729.68774.a1]

CHAPTER 6

Reproductive Biomarkers as Early Indicators for Assessing Environmental Health Risk

Luigi Montano[*]

Chief of Andrology Unit and Lifestyle Medicine, Local Health Authority (ASL) Salerno, EcoFoodFertility Project Coordination Unit, Oliveto Citra (SA), Italy

Abstract: The evaluation of exposure in association with information on the inherent toxicity of the chemical (that is, the expected response to a given level of exposure) plays a critical role to predict the probability, nature, and magnitude of the adverse health effects. The epidemiological findings, the results on cancers and other chronic diseases with a long latency are a weak tool to reduce the risk for the current and next generation. An important issue is the exposure time into health risk assessment/management, especially in highly polluted areas, where health problems increase. In this regard, the environment and health aspects must become, as a matter of urgency, an international priority, both in terms of policy and resource allocation. The use of reproductive biomarkers for early risk detection is introduced by the EcoFoodFertility research project in one of the areas with the highest environmental impact in Europe, "Land of Fires" in Southern Italy. This area, a symbol of the ecological crisis, represents a possible new methodological approach in public health. This chapter aims to explain how biomarkers of reproductive health could be exploited as early flags of environmental pressure and enhanced risk of chronic adverse effects on health. In particular, human semen seems to be a sensitive source of biomarkers, giving information on biologically active exposures, and it can be very useful for preventive health surveillance programs, especially in environmental risk areas. This approach appears very promising, above all, in young people (maximum fertile age:18-35 years), considering the possibility to reduce the chronic-degenerative diseases in future adults. In this context, many scientific findings are increasingly about the association between pollution and fertility problems and therefore, the safeguard of germ cells is a new challenge to reduce the burden of epigenetically transmitted diseases.

Keywords: DOHaD, Ecofoodfertility, Environmental Marker, Endocrine Disruptors, Environmental Health, Epigenetic, Health Marker, Human Semen, Land of Fires, POHaD, Public Health, Pollution, Reproductive Health, Semen Quality, Sperm Epigenome.

---

[*] **Corresponding author Luigi Montano:** UroAndrologist, Chief of Andrology Unit and Lifestyle Medicine, Local Health Authority (ASL) Salerno,EcoFoodFertility Project Coordination Unit, Oliveto Citra (SA), Italy; Tel: 00393339433861; E-mails: l.montano@aslsalerno.it and luigimontano@gmail.com

INTRODUCTION

The World Health Organization (WHO) and the Organisation for Economic Co-operation and Development OECD, (https://www.oecd.org/china/air-pollution-to-cause-6-9-million-premature-deaths-and-cost-1-gdp-by-2060.htm) calculated about 600,000 premature deaths and diseases caused by indoor and outdoor air pollution in the European Union [1]. Furthermore, many findings point out the responsibility of human activities on climate change which is becoming very dramatic. Naturally, human lifestyle plays a central role in the onset of many chronic diseases, as reported in the European Code against Cancer, although the effects of chemical and physical pollutants actually represent the most important threat to public health whose transgenerational effects should cause considerable concern to policymakers. In fact, most of the outdoor and indoor contaminants exert their action as endocrine disruptors by altering cellular signals and also, inducing oxidative stress (excess of free radicals of oxygen not balanced by the presence of reductive activity). The oxidative stress is able to damage at biomolecular level DNA, proteins, lipids, which, if not properly repaired, trigger a significant inflammatory response and, in turn, induces neoplastic transformation. In this regard, the imbalance between antioxidant defenses and detoxification processes produces the onset of oxidative stress diseases in humans, due to increased susceptibility of the organism to pollutants. The pro-oxidant activity of particulate matter (PM) [2] polycyclic aromatic hydrocarbons (PAH) [3] on human health has been demonstrated in clinical data, while the harmful effects, induced by toxic heavy metals or organophosphate pesticides, has been demonstrated in animal studies [4]. Exposure to environmental toxicants contributes to the increase in the production of free oxygen radicals, while the nuclear factor erythroid 2, like 2 (Nrf2), is able to modulate the antioxidant response through the cellular antioxidant system, the reduced glutathione (GSH), and the activity of the Glutathione reductase (GSR) [5, 6]. Its reduced activity has been associated with the onset of pathologies, elevated susceptibility to negative effects due to pollutants and, additionally, with overexpression of p53 [7]. Other biomarkers for pollution are metallothionein proteins, which are used as a contaminant-specific indicator of metal exposure [8]. They are inducible proteins and the accumulation of heavy metal cations within the cells enhances metallothionein gene transcription by stimulating their synthesis. The metall-othionein messenger RNA is translated by the cytosolic free ribosome; it leads to an increase in apo-metallothionein that rapidly reacts with the free metal cations that are present in the cytosol [8]. These proteins can protect cell structures from non-specific interactions with heavy metal cations and to detoxify the metal excess penetrating into the cell. In this regard, metallothionein proteins are usually considered to be important specific biomarkers that detect an organism's response to inorganic pollutants such as cadmium (Cd), mercury (Hg), copper (Cu), and zinc (Zn) that are present in the environment. The toxicity of heavy metals may be attributed to the binding of metals to sulfhydryl groups in proteins such as glutathione (GSH), resulting in an inhibition of activity, interference with structure, or displacement of an essential metal element leading to deficiency effects [9]. Repairing stress-damaged proteins and chelation of metals involving heat shock proteins and metallothionein is thus recognized as a potential mechanism of metal detoxification. Although mechanisms by which heavy metals interact have not been clearly elucidated, a number of biomolecules, including GSH metallothionein, and heat shock proteins have been predominantly recognized as major interactive mediators when evaluating interactions based on metal mixture exposure [9]. The striking feature of metallothionein is the inducibility of MT-1 and MT-2 genes by different agents and conditions. The regulation of metallothionein biosynthesis happens primarily at the level of transcription. The MT-1 and MT-2 genes in higher species are rapidly induced *in vitro* and *in vivo* by a variety of stimuli including metals, hormones, cytokines, oxidants, stress and irradiation [10]. Owing to their induction by a variety of stimuli, metallothionein proteins are considered to be valid biomarkers in the medical and environmental fields [11]. Numerous studies have demonstrated that changes in MT expression are associated with the process of carcinogenesis and cancer progression. However, the expression of MTs is not universal in all human cancers. Previous studies have shown that MT expression is upregulated in breast cancer, nasopharyngeal cancer, ovarian cancer, urinary bladder cancer, and melanoma [12 - 16], while in other cancers, such as hepatocellular carcinoma, prostate cancer, and papillary thyroid carcinoma, MT expression is downregulated [17 - 21]. Moreover, tumors arise from the interaction between intrinsic factors (constitutional variants and genetic and/or molecular alterations) and extrinsic factors (levels of environmental carcinogens and pro-carcinogens with which one comes into contact during life). In particular, the sequential acquisition and the relative accumulation of molecular alterations involved in tumor development and progression seem to be influenced by the presence of different genetic susceptibility factors (mutations and gene polymorphisms) in association with diet, lifestyle and pollution. In addition, systemic chronic inflammation (SCI), triggered by environmental factors, is a critical etiological element of many non-communicable diseases (NCDs), when the reduced immunological defenses can lead to neurodegenerative, cardiovascular, metabolic diseases and cancer [22]. Furthermore, systemic chronic inflammation during pregnancy can modify the expression of key genes in the foetus leading to an increased risk for NCDs. Such diseases can persist in the lifetime of new individuals and potentially influence future generations. Furthermore, very recent studies have reported that the modifications of germ cells epigenome, induced by lifestyles and exposure to contaminants of parents, may also partly explain the growth of cancer incidence in children and adolescents in Europe from 1991 to 2010, particularly in Southern Europe [23]. It is therefore important to promote tools to reduce exposures and health impact through risk awareness and health-promoting behaviors and practices, and also to prioritize the development of new methods for the early prevention and diagnosis of SCI and the use of clinically validated biomarkers for early risk detection. Essentially, if the classical epidemiology assess health risk by "counting" the final outcomes of health damage (mortality, incidence, hospitalization for diseases, tumors, *etc.* through registers) with a long latency, it's necessary to support risk reduction for the next generations, evaluating the earliest signs of functional or structural modification, before occurring the clinical damage. In this scenario, it becomes necessary to evaluate the "time" dimension into health risk assessment/management with a strong focus on the future of communities from the ethical as well as scientific standpoints [24].

This need is even more true in areas where environmental pressure is higher, *i.e.* where there are actually more unfavorable health indices with a higher incidence of chronic-degenerative diseases such as oncological ones [25], index of overall health more unfavorable [26] and reduced fertility [27 - 32]. One of those areas where concentration of pollutants with well-known detrimental biological effects has reached alarming levels is "Land of fires" (Fig. **1**). This is the designation under which sadly today is connoted the area between the provinces of Naples and Caserta, the heart of the "Campania Felix", characterized by a multiplicity of sources of pollution (illegal disposal of urban, toxic and industrial wastes, dumping practices, traffic, intensive agriculture), widespread on high territorial extension with 2,5 million inhabitants, making this area the symbol of the ecological crisis in Italy, the echo of which crossed national borders. This is one of the best-defined populations in terms of environmental pollution in Europe and an ideal area for exposome studies. Several studies highlighted [33 - 36] a significant increase in mortality in young people in particular, a high incidence of some hormone-sensitive tumors, both in males (testis and prostate) and females (breast and uterus) [37], although evidence is not conclusive. Other studies showed that increased levels in heavy metals (Hg, As, Cr and Ni, Cd and Pb) caused acute and chronic toxicity (reproductive and endocrine systems, immunosuppression) in animal samples [38].

Exactly from this area, in a perspective of primary and pre-primary prevention was born the Ecofoodfertility project that focuses precisely on biomarkers to better and first assess the effects of pollution and provide early and predictive health risk [39].

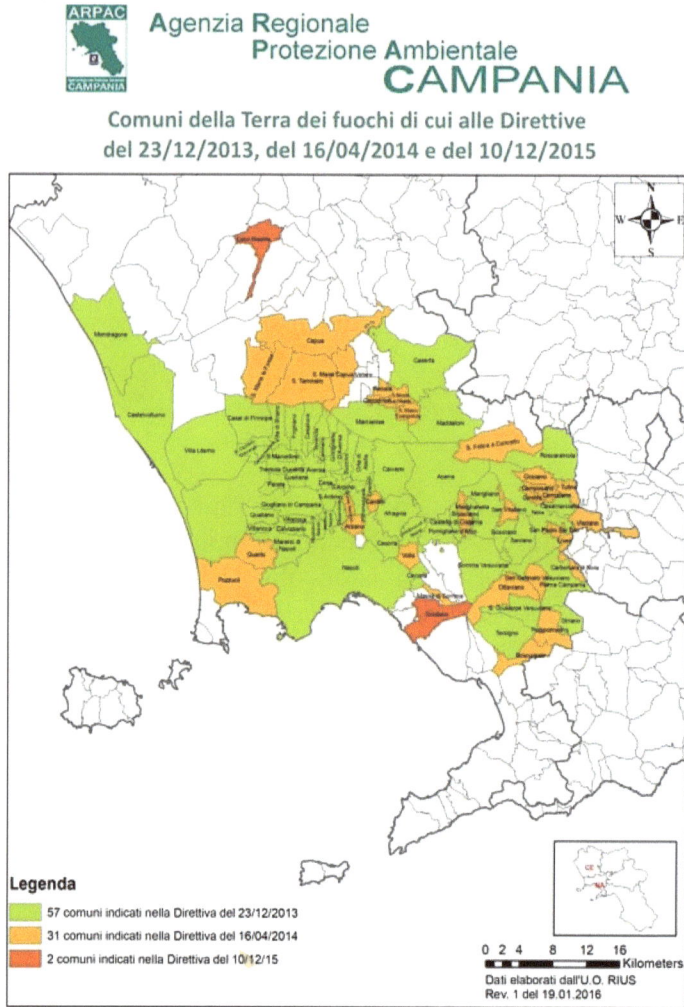

Fig. (1). Map of the municipalities of "Land of Fires" (http://www.salute.gov.it/portale/news/ARPAC, 03/08/ 2017 e http://www.arpacampania.it/aria).

In fact, this project introducing an integrated and direct approach to address assessment of the "*early health risk*", through a systematic biomonitoring study for several contaminants (Metals, Dioxins, PCBs, Polycyclic Aromatic Hydrocarbons, Bisphenol A, phthalates, Parabens, pesticides) and an analysis of biomarkers of exposure, damage/effect and susceptibility in blood and semen on selected and homogeneous groups by age and lifestyle habits of healthy males in the different areas of Campania, Italy and Europe. The aim is to see whether there are differences between groups, and to measure, with data integration, the earliest signs of functional or structural modification, before clinical damage manifests itself and, therefore, have an estimate of possible differences of early health risk, evaluating the environmental weight for each territorial context in relation to individual variables related to eating habits and lifestyles. The study takes into account the functional-organ systems which earlier than others suffer from the stress conditions, such as endocrine-metabolic system and, especially, the male reproductive system (Sentinel Organ), as an early model of study and general health check detector.

The Aims:

a) developing a better understanding of the effects of environmental pollutants on human health in areas with different environmental pressures:

b) qualifying human semen as an early and sensitive *"Environment and Health marker"*;

c) proposing *"human seminal model"* for early detection and prevention of environmental health risks, useful in innovative programs on health surveillance;

d) identifying a dietary treatment that can reduce pollutant bioaccumulation and improve human semen quality in healthy men who live in polluted areas.

This research represents an example of citizen action supported by several environmental groups of the territory and involving different public research institutions. Furthermore, such project represents a response to an environmental and health crisis and also to better understand the contradicting findings and media reports. Overall, this little clarity generated a distrust in the public healthcare system as well as anxiety within the population. Currently the project model is already contributing to the development of new studies in both regional and national institutions in order to build a large network of action in other contaminated areas of Italy. Several until now are the congress communications and publications on the differences found between homogeneous subjects in different environmental contexts starting from the first study published in 2016 [39] that highlighted in 110 healthy male subjects, no smokers, homogeneous for age and lifestyle an increase of several metals in blood (Al, Mn, Cr, Mg, Li, Co and Ca) and seminal liquid (Cr and Zn), comparing residents of the Land of Fires with subjects living in an area with low environmental impact in the province of Salerno ("Valle del Sele").

THE CHALLENGES OF CONTAMINATED SITES REMEDIATION

One of the major environmental problems facing the world is cleaning up hazardous waste sites from past or current industrial activities. National governments, as well as many companies, are interested in cleanup of contaminated lands for productive land uses, as well as the protection of human health and the environment. Cleaning sites for future development is a major issue due to their world-wide occurrence [40]. The importance of preserving natural areas plays a key role in a realization of a sustainable future considering a high increase of industrialization, urbanization, and globalizations. In this context, science-based decisions are required about how to remediate, cleanup levels, to protect both human health and the environment for long-term [41 - 48]. A range of stakeholders in association with ecologists should be involved early and often in cleanup decisions [49 - 51]. Science-based decisions can only made when there is knowledge exchange among all parties [47]. Furthermore, the improving the health and well-being of people passes pass through an ever-increasing awareness about the role of ecological resources and services, eco-cultural resources, and other aspects of ecosystems [52, 53]. The integration between remediation and restoration with ecological, ecocultural, and cultural protection and enhancement [54 - 57] is necessary to protect the health of communities [53, 58]. In this regard, the remediation of hazardous waste site should be performed in a sustainable manner [51, 59 - 61]. In many places, land is in short supply, and considered more valuable (often due to increasing populations or affluence). Today, the public pressure is increasing to remediate the contaminated lands in order to to preserve human health and the environment [62], to reuse these lands and to increase resiliency [63 - 65]. Moreover, it is important to focus on ecosystem restoration [66, 67] through sustainable methods of interdiction before and during remediation on contaminated lands to protect the existing ecological resources. In their study, Wagner *et al.* pointed out the importance of establishing a set of achievable restoration goals for contaminated sites in concert with establishment of the remediation goals themselves. Naturally, it is essential to include a range of technical expertise in association with well-informed decision making leading to agree upon future land uses. Furthermore, it is important to provide a framework for reducing the harm to the existing ecological resources on contaminated sites from remediation, including specific suggestions for on-site remediation workers as shown in Table **1**.

Table 1. Major decision points where intervention can improve the process to protect and enhance ecological resources.

Major Stage or Phase	Steps or Considerations	Examples of Points of Interdiction for Protecting or Enhancing Eco-resources
Site identification	Recognition of a site needing remediation	Maintain vigilance about local and regional contaminated sites.
		Maintain vigilance about historic industries or activities that may require cleanup.

(Table 1) cont.....

Major Stage or Phase	Steps or Considerations	Examples of Points of Interdiction for Protecting or Enhancing Eco-resources
		Keep involved at every decision point once a contaminated site is identified.
		Track local government-recognized sites, such as EPA's superfund sites in US (EPA, 2018) [1].
		Identify land owner or responsible party.
		Be familiar with federal and state laws (*e.g.* CERCLA, RCRA) (*e.g.* EPA, 1990, 1991; ATSDR, 2007, 2013; Azam, 2016) [2 - 6].
		Identify the site boundary and the point of compliance for regulatory considerations (WSDE, 2007; Sample *et al.*, 2014) [7, 8].
		Be familiar with the local ecology, including habitats (eco-region, Omernik, 2004) [9].
Problem identification	Be part of the original problem identification	Be familiar with discussions concerning whether a given site is being considered for remediation, and note the importance of protecting human and ecological health (Reagan, 2006; Sandifer *et al.*, 2015) [10, 11] .
		Be sure sustainability is a consideration for the postremediation landscape (Ellis and Hadley, 2009; Holland, 2011; Owsianiak *et al.*, 2013) [12, 14] .
	Find information on contaminants of concern	Find information on whether the known contaminants pose a risk to eco-resources (ATSDR, 2007, 2013, EPA, 1997; Bartell *et al.*, 2002;) [4, 5, 15, 16] both in aquatic and terrestrial systems (O'Halloran, 2006) [17].
	Consider the site and who might have a stake in its management	Identify interested and affected parties and all stakeholders (government agencies, regulators, natural resource trustees, site neighbors, advisory boards) to be lobbied on behalf of ecological resources (Abbotts and Takaro 2005; NRC, 2008; Paavola and Hubacek, 2013) [18 - 20]. Develop collaborative partnerships (Chaffin *et al.*, 2015) [21], including with environmental justice communities (EPA, 2009; Burger, 2011) [22, 23].
Regional ecological resources	Determine regional and local ecological type	Determine the value of the ecological resources on the contaminated site compared to the local region (Bingham *et al.*, 1995; Costanza *et al.*, 1997, 2014) [24 - 26] . The U.S. is divided into eco-regions, useful for understanding local ecotype (Omernik, 2004) [7].
		Determine if the region has any state or federal endangered/threatened species (ESA, 1973) [27] .

(Table 1) cont.....

Major Stage or Phase	Steps or Considerations	Examples of Points of Interdiction for Protecting or Enhancing Eco-resources
Local land use .	Determine how land use decisions are made	Determine if there are zoning or other land use restrictions unrelated to residual contamination (Gochfeld *et al.*, 2015) [28].
		Search information on future land uses for the site, including helping designate some key lands for ecological/conservation use on site (Leitao and Ahern, 2002; Whicker *et al.*, 2004) [29, 30] .
Remediation goals	Determine whether owner or federal/state government are planning on cleanup	Determine whether and when remediation may be in the planning phase, or whether a decision has been made about the type of remediation.
Remediation options	Determine the remediation alternatives being considered	Obtain necessary documents (Remedial Investigation/Feasibility Study, Record of Decision) to allow evaluation of each option with respect to functional effects of remediation (EPA, 1990; Burger *et al.*, 2016) [2, 31].
Global issues	Identify what global issues may be important to the site	Consider how global issues (demographics) might affect specific contaminated sites and their ecological resources (*e.g.* population increases, shifts in where people live) (check census data for trends).
		Consider effect of climate change (and sea level rise) on local ecology (IPCC, 2014, 2018) [32, 33], and initiating events (*e.g.* earthquakes, fires; Burger and Gochfeld, 2016) [34].

In conclusion, in the case of contaminated sites, environmental epidemiology can contribute to better understand the approaches for sustainable development practices. Furthermore, these practices allow a better understanding of environmental contaminations and health effects pathways in order to propose policy options for environmental risk prevention, remediation and management.

BIOMARKERS FOR THE EVALUATION OF HEALTH RISK

Biological markers represent a substance used as an indicator of a biological state. They are measured and evaluated as indicators of normal biological and pathogenic processes, or pharmacologic responses to a therapeutic intervention [68, 69]. Moreover, these biomarkers might identify the risk of the presence of disease [70, 71]. In the mid-80's, many studies began to demonstrate that the environmental contaminants caused various damages to organisms at various levels of hierarchical organization, from the molecule to the community. Every living organism, plant or animal that can provide information on the level of contamination in a designated area, is called "Sentinel Organism". In this perspective, biomarkers or stress indexes, defined as a contaminant-induced alteration, at molecular or cellular level of a structure or function, can be highlighted and quantified in a sentinel organism. When a toxic compound penetrates an organism, it can cause a number of alterations at molecular levels damage, up to alterations in organisms, and in turn, in populations or community. Therefore, the principal aim is the identification of the primary target of the action of a polluting compound. Primary toxicity of a contaminant is exercised at the molecular level, causing changes in enzyme activities, DNA-level changes, *etc.* Subsequently, the effects can be detected with a cascading mechanism to the next levels of hierarchical organization, organelle, cell, tissue, organ and organism up to the population level. Simultaneously, every organism is able to develop adaptive responses to the chemical stress in order to restore a normal homeostasis state. In particular, at beginning, the homeostatic responses decrease the toxic effect of the polluted compounds by activating multi-enzymatic systems designed to detoxify completely or partially the organism. In this regard, the different systems with whom the organism responds to the chemical insults, represent the "potential biomarkers" to use in environmental monitoring programs. In this way. it could be possible to predict long-term adverse effects such as carcinogenesis, pathological alterations, the reduction of reproductive abilities and the mortality in a population through the study of immediate responses.

Biomarkers' Interpretation

Environmental pollutants generate harmful conditions for living organisms, including humans. In this case, the living organisms are exposed to different chemical compounds that interact actively with each other resulting in a cumulative or synergistic response in the individual. In addition to anthropogenic factors, the body of organism is influenced by many natural chemical and physical factors (temperature, salinity, oxygen, *etc.*), ecological and physiological (hormonal status, nutritional status, age, *etc.*), that are involved in the metabolic responses of the organism. At this point, it is clear that in nature every organism faces with a "family" of dose-effect reactions, that correspond to the different interactions between pollutants and environmental stresses. Then, the interpretation of biomarkers is based on a new conceptual paradigm, which can be summarized as follows: the use of biomarkers do not evaluate the quantitative levels of the toxic compound in the body, but the determination of "health status" of organism and in its transition from a normal state of homeostasis to the disease. When a body is subjected to the effect of toxic compounds, a series of biochemical and physiological protective mechanisms that tend to bring the system back to a state of normal homeostasis are triggered; if the exposure continues or increases, the compensatory processes become inefficient, the repair systems start to function . When those mechanisms are no longer sufficient, there is a stage of visible manifestations of toxic effects with different responses up to carcinogenesis and the death. Since the use of the different biomarkers with subsequent response times enable to identify the state the population is in (homeostasis, compensatory responses, responses for shelter, illness), an assessment of "risk level" can be done.

Biomarkers' Benefits

The contamination effects assessment of a community that lives in an ecosystem is a problem of difficult resolution for several reasons, among which are included the following:

• There are several possible routes in the organism for the pollutants intake;

• The contaminants have a different bioavailability according to environmental spheres in which they are found;

• Individuals are typically subjected to a mixture of contaminants: these substances can produce different biochemical and toxicological interactions between each others;

• There is a very long latency period before the alterations at the level of populations and communities are evident.

The bioindication qualitatively assesses biotic responses to environmental stress. Hereafter, the term "bioindication" is used as a collective term to refer to all terms relating to the detection of biotic responses to environmental stress. The biomonitoring provides additional information with respect to chemical or toxicological monitoring. In this regard, environmental chemical analysis can be incomplete for environmental quality studies, since they provide an accurate quantitative and qualitative point of view on the distribution of pollutants in the various sectors, but they do not include the contaminants interaction on the organism. Given the growing number of pollutants, these analyses result extremely costly in terms of men and means. These problems can be solved with the use of biomarkers in sentinel organisms. These two methods bioindication and biomonitoring provide an integrated response of the overall exposure of sentinel species to toxic compounds by evaluating both the different pathways of the pollutants and the exposure within a specified interval time. They provide an immediate response to the exposure to toxic substances. This allows to predict the long-term negative effect. There are, however, also natural factors that affect the physiological state of the organism and therefore alter in a certain way the signal given by the indexes of stress. These disturbing factors such as hormonal status, age and sex of the body can affect the biochemical reactions, and this must be taken into account.

Biomarkers' Classification in the Interaction Between Organism and Contaminant

A first subdivision can concern the increased levels of interaction between the contaminant and the immune system:

• Biomarkers of exposure represent an early biochemical response after exposure of an individual or organism to a contaminant For this purpose, stress indexes can be used, such as the enzymatic activities of mixed-function monooxygenase (MFO), or the metallothionein as exposure signals to organic chlorine compounds and heavy metals, the acetylcholinesterase inhibition as a result of exposure to organic phosphor and carbamate-based insecticides or the quantification of DNA adducts derived from polycyclic aromatic hydrocarbons (PAHs):

• Biomarkers of effect: they indicate how an organism, a population or community are subject to toxicological effects from one or more pollutants and in particular are able to measure early pato-physiological variations (subclinical signs), rather than a full-blown clinical disease.

• Biomarkers of susceptibility (genetic polymorphism) predict the individual variability of disease risk in relation to environmental exposure. Notably, the telomere shortening is able to identify early damage in the DNA and is scientifically validated as early markers of risk of environmental diseases (such as cancer and atherosclerosis). Telomere Length - terminal region of each chromosome is considered a marker of biological age. Indeed, it decreases (physiologically) with age but also, and especially, pathologically upon exposure to environmental insults. The shortening of telomere's length increases genomic instability, which in turn facilitates the accumulation of genetic alterations that may contribute to the onset of diseases.

Polymorphisms of genes involved in DNA mechanisms of repair and detoxification can create the basis for interindividual differences in the risk of developing clinical events associated with environmental exposure factors. The analysis of functional polymorphisms of genes involved in metabolic detoxification mechanisms (*i.e.* for benzo(a)pyrene and the genes of CYP1A1, CYP2E1, GSTM1 and GSTT1), allows to predict individual differences in response to the effects resulting from exposure to environmental factors.

Similarly, DNA damage repair mechanisms are essential to maintain the integrity of the genome that can be weakened by environmental toxins. Functional polymorphisms of genes involved in DNA repair mechanisms (*i.e.* XRCC3, hOGG1) are then responsible for a minor restorative capacity.

Specificity of the Response

Biomarkers can be also divided according to their "specificity" responsiveness of pollutants in:

• Specific Biomarkers: are molecular and biochemical responses of an organism to a specific class of contaminants (such as metallothionein in response to metal pollution). In this case the defense response is extremely specific and clearly indicates the class of substances responsible for the contamination.

• General Biomarkers are those of the organism, at the molecular, cellular and physiological level, that cannot be traced to a specific class of pollutants but represent the general condition of the body's stress (certain immune disorders, DNA damage, somatic indices, Lysosomal membrane stability *etc.*).

Biomarkers as a Diagnostic Tool

In biomonitoring analysis, it is possible to detect the presence of particular pollutants with specific biomarkers or a suffering state of organisms through general biomarkers. There are three levels of hierarchy in which biomarkers can be applied in such kind of analysis:

1. The first stage is the identification of chemical hazard present in the contaminant mixture. Such identification is possible by using general biomarkers, and in this case, the presence or absence of a chemical hazard can be detected.

2. The second stage is the hazard assessment, since the potential pollutants are known. At this stage, the use of specific biomarker allows us to identify classes of contaminants, the extent and severity of contamination area.

3. The last stage is the prediction of risk, in this case, biomarkers analysis may provide the potential long-term negative consequences at the level of population and community.

FROM BIOMARKERS TO SENTINEL ORGANS TO DETECT EARLIEST RISK INDEXES

Information on exposure levels to both chemical and physical contaminants can play a critical role to evaluate environmental and professional risks. Consequently, such information can estimate the biological risk of potential hazardous chemicals in terms of exposure to some chemical and/or physical stress. Environmental monitoring in a certain spatial and temporal context enables to assess the exposure through the measure of concentration of some compounds and their metabolites. In this context, the analysis of the biological risk assessment for the community is very complex being associated with numerous environmental variables. Besides, the knowledge of pollutants concentration in the environment and their seasonal variability, is essential, although it is difficult. In this way, the biomonitoring might help us to reduce the biological risk. In this scenario, it needs to consider the individual lifestyle, in the broadest sense (including environment, habits, use of cosmetics, use of plastic bottles, using the pc, wireless internet and much more). Then, after characterization of chemical and physical factors at working and individual level, it is possible to evaluate the absorption pathway of these chemicals or their metabolites (such absorption varies from individual to individual and from variable conditions for the same individual), through the analysis of the biological fluids or the tissues. In this context, when we are able to measure these values in fluids or tissues of individuals, we can ascertain the biological effect (measurement of biomarkers of effect) of individual response (markers of genetic susceptibility: polymorphisms *etc.*). The combination of environmental data and the chemical substance concentrations found in the fluids or tissues, it will allow us to better understand the passage from the environment to the organism. Therefore, the measures of these substances and the resulting biological effect will permit us to establish a close relationship between exposure dose and dose effect. Furthermore, in order to adopt an effective primary prevention strategy, the identification of exposure pathway, most sensitive system-functional organ to the exposure and, simultaneously, the more sensitive biological medium, which, before others, is predictive of future damage, represent the strategic objective that can help detect the earliest clinical risk indices.

THE IMPACT OF CONTAMINANTS ON THE MALE REPRODUCTIVE SYSTEM: THE SENTINEL ORGANS

Over the past 60 years, multiple studies showed a decreased sperm concentration in many industrialized countries, and the incidence of male infertility has undergone a drastic increase (from 7-8% in the 60's to 20%-30% today). In addition, demographic surveys reported a constant reduction in birth rates in all European countries in early 50's. In 1992, Carlsen, in a meta-analysis conducted on 61 European studies, examined samples of donors' semen analyses from 1934 to 1990, and observed a progressive worsening of qualitative and quantitative characteristics of semen (from 113 Mil/ml in 1940 to 66 Mil/ml in 1990 and a decline of 19% of the volume ejaculated) [72, 73]. In another study, released in December 2012 on Human Reproduction, French researchers of the Institut de Veille Sanitaire, in a comparison between the semen analysis of 26,600 men, collected between 1989 and 2005, showed a further reduction of the average concentration of spermatozoa [74]. Lastly a metanalysis of Levine reported in several western countries (Europe, U.S.A, Canada and New Zeland) a decline of total sperm counts by 59.3% between 1971 and 2011 [75]. Today, the increase in couple infertility is becoming a priority for public health, because, in addition to psychological distress and high economic costs, it is possible to observe evidence of diseases associated with poor semen quality [76, 77], testicular cancer [78 - 80], other types of cancer [81 - 83], several comorbidities [84 - 88] cross-generational effects [89, 90] and shorter life expectancy [91]. Over the past 40 years, there have been dramatic changes in behavioral factors and lifestyles that may potentially induce alterations in seminal parameters and thus reduce male fertility, including the introduction and rapid growth of cell phones' use, the large increase in the consumption of opiates and marijuana, the increase in the consumption of cigarettes, increasing obesity and physical inactivity [92]. Furthermore, environmental and chemical contaminants in the workplace are recognized as major risk factors for male infertility in both epidemiological and experimental studies [93 - 97]. The incidence of genitourinary tract malformations and reduced sperm quality is higher in people living in areas with a high rate of pollution or in individuals exposed for professional reasons. The latter, in fact, show a reduction of concentration, motility, morphology and/or sperm DNA damage. In this regard, various toxicology studies in animal models report damage on the germline and the DNA and, additionally, as a result of exposure to environmental xenobiotics and toxicants during the foetal development and early postnatal life, causes sterility in next generations as well as congenital malformations and degenerative diseases. There are therefore several findings that show how ubiquitous presence of chemicals in the environment and in food can actually be the root cause of this reduction of the different quantitative aspects of human semen. Furthermore, the harmful environmental factors can play a critical role in the increased incidence of testicular dyskinesias in the recent years. A milestone in understanding of such pathogenesis was the discovery of cancer cells presence *in situ* in adults, affected by testicular cancer [98]. Specifically, germ cells (gonocyte type) were transformed into cancer cells and could not differentiate into spermatogonia during the foetal period [99]. The incidence of testicular cancer is increasing in all regions with increased variability between populations. It appears that the genetic contribution is important since African Americans have a significantly lower incidence compared to Caucasians living in the same region of the USA. Families with two or more cases of testicular cancer have been reported, suggesting that there are genes responsible for susceptibility to cancer [100]. But not all genetic differences are explainable: there are a big difference even between two populations of Northern Europe that have a similar social and economic structure: Denmark and Finland. The Denmark has the highest incidence of testicular cancer, while Finland has the lowest incidence, and it has been argued that environmental factors can be the cause of this difference [101]. In particular, it has been found, in Denmark, that the sperm quality of the general population was lower, and the incidence of cryptorchidism higher than in Finland, supporting the hypothesis that these disorders could be epidemiologically linked to each other. In this context the testicular dysgenesis would also be related to genetic factors such as susceptibility to endocrine "disruptors" [30]. Few studies on humans have found associations/correlations between endocrine "disruptors" and the different components of testicular dysgenesis. However, for ethical reasons, on the human being, it is difficult to establish a causal relationship, making the animal model recently developed an important tool to investigate the pathogenesis. Clinically, the most common manifestation is a reduced sperm concentration, while the most severe form can include an increased risk of testicular cancer [102]. Concerns about the potential risk of endocrine "disruptors" on male reproductive endocrine function have been underlined in studies that showed a downward trend in the semen quality, as well as an increased incidence of diseases and testicular cancer. Certain chemicals including phthalates, pesticides, and polychlorinated biphenyls (PCBs) [103] and additional classes of environmental pollutants require further studies on their relation with the decay of human semen quality [104]. The predictive value of substances having a toxic effect on the male reproductive tract has been very discussed in an important Congress, named "NESOT 2012 Annual Meeting, Translational Biomarkers in Toxicology" in Newport, Rhode Island, laying the foundations for a scientific research [105]. More strikingly, especially in industrialized countries, the reduction of birth rate, is not dependent on the increased use of contraceptives and the number of abortions, but due to a reduction in fertility, which presents differences in areas within the same country or even in the same region. In For example, in Italy, a research performed on 10,000 men (average age 29 years) conducted by Prof. Fabrizio Menchini Fabris, of Pisa University, has shown that there is a great national territorial variability, linked to environmental factors, food and lifestyle. In Italy, the metropolitan urban areas (Rome, Milan, Naples and Palermo that show a worsening ranging from 10% to 17% with a negative record for Naples) are more effected than urban centers or rural areas, especially those with a population of less than 20,000 inhabitants [27]. In this case, heavy metal pollution, particulate matter, known and unknown pollutants, combined with the stress of urban life, the use of alcohol and drugs and the drastic reduction of the hours of sleep, degrades the environment where sperm cells mature. The first systematic study on the association between environmental pollution and human reproduction was conducted in the Czech Republic with the research program "Teplice". In particular, such study showed an increased fragmentation of chromatin, an elevated abnormal shapes and an increase in the rate of sperm aneuploidies, which correlated with the increased atmospheric pollution (higher concentration in the air of polycyclic aromatic hydrocarbons PAH and PM10), and occurs more in winter, compared to summer. DNA adducts produced by PAH in blood and placenta were found in the inhabitants of the same region. The authors of this study speculate that DNA-PAH adducts can also form in sperm of exposed individuals and be responsible for the increased DNA fragmentation observed [106].

EPIGENETIC DAMAGES OF ENVIRONMENTAL CONTAMINANTS: THE SPERM EPIGENOME

In recent years, new acquisitions have allowed to better understand gene expression regulation and epigenetic mechanisms. Many studies of gene–environment interactions provided more insight into the biological processes. Furthermore, the epigenetic studies showed specific gene expression changes without any change in their sequence. Therefore, certain environmental insults and epigenetic alterations can generate genetic variants that make vulnerable the organism. Today, the sperm epigenome studies found the relationship between altered epigenetic mechanisms and environmental exposures. Such mechanisms are represented by DNA methylation, histone modifications and noncoding microRNAs [107]. Furthermore, several studies demonstrated he correlation between sperm DNA methylation and idiopathic male infertility [108 - 111]. Many authors showed that DNA hypermethylation of gene promoters (like MTHFR, PAX8, NTF3, SFN and others) was involved in male infertility. On the contrary, hypomethylation of other genes the check zone IGF2/H19 1 (ICR1), was observed in patients with both lower sperm concentration and motility when compared to controls with normal sperm kinetics [112 - 118]. Nuclear condensation in the spermatozoa is another sensitive stress event, that is able to induce genetic and epigenetic alterations. During this phase, 85% of histones (rich in lysine) are bound to DNA, and they are replaced with proteins of transition and arginine-rich proteins: the protamine [119, 120]. Such proteins are linked to DNA helix grooves, and they wrap themselves tightly around the strands of DNA (about 50 kb of DNA and protamine), in order to form highly organized loops. In addition, the compaction and stabilization of the spermatozoa's nucleus occurs through the intramolecular disulfide bonds between cysteine-rich protamine. Such associations are an extreme nuclear condensation that decreases about 10% of the size of the nucleus. The key protein involved in this process is the BRDT (Bromo Domain Testis-specific) which promotes nuclear remodeling allowing the transition between a histone chromatin organization (that is somatic, while the protamine is typical of the mature sperm). This compaction is useful to protect the sperm genome from external stress. Different factors (such as physiological and environmental stress, genetic mutations and chromosomal abnormalities) can interfere with the mechanisms of spermatogenesis [121] and induce alterations in abnormal chromatin structure which is incompatible with fertility. The presence of the genomic material defects in mature sperm may be responsible of packing defects (defective replacements of histones-protamines), defects in the maturation of the nucleus, DNA fragmentation defects (that is, single or double strand breaks), sperm DNA integrity defects or chromosomal aneuploidy and changes in gene expression (epigenetic modifications). In particular, many data support this hypothesis that the DNA is not homogeneously rich of protamine in the mature spermatozoa of mammals [122]. The transcription of genes can be affected by defects in the action of protamine. For example, the deregulated protamine action process in mice causes premature chromatin condensation, interruption of the transcription, and failure of spermatogenesis [123]. It is reported that the 10%-15% of original histone content is conserved in the human sperm's nucleus and distributes in the genome in heterogenous manner [124]. A study of the entire genome of seven infertile patients showed that a random process of protamine action was occurred in five of the seven infertile men when compared with normal fertile men [125]. These specific errors in the epigenetic control process can happen at every stage of the spermatogenesis, and cause damage of male fertility and embryonic development. In fact, the expression of specific genes involved in the early stages of spermatogenesis can be altered by epigenetic changes, that lead to a decreased efficiency of the process. During meiotic stage, these changes are able to induce double strand breaks or chromosomal nondisjunction. Finally, defects in the histone-protamine transition and/or the histone removal difficulties and the protamine replacement errors may be involved in epigenetic changes during the spermiogenesis [121]. Taken together, these results suggest that the male infertility (such as alterations in sperm count or morphology, DNA fragmentation chromosomal, aneuploidy, alterations in the chromatin density) could be dependent on epigenetic mechanisms that occur at different stages of spermatogenesis. Furthermore, the microRNAs (miRNA) and the posttranscriptional regulation can be involved in these processes [126 - 128]. Specifically, miRNAs are expressed during spermatogenesis and, they control every phase of the male germ cell differentiation. Many findings derived from genetically altered rat models found that the specific miRNAs regulated the development of a normal spermatogenesis [129]. At the end, clinical studies showed that spermatozoa derived from patients with sperm alterations had altered miRNA profiles [130, 131]. In a study, Wang *et al.* examined a pool of sperm samples derived from fertile and infertile men, and the authors found several alterations in miRNA profiles both in azoospermia and asthenozoospermia conditions [132]. In particular, the patients with azoospermia had a low level of seven miRNAs, while the patient with asthenozoospermia had high level of same miRNAs, compared to fertile subjects considered as case-control. For this reason, the authors proposed a molecular diagnostic value for male infertility of these miRNAs. In this regard, they evaluated the expression pattern of miR-I9B and let-7 bis in patients with idiopathic infertility, azoospermia or non-obstructive oligozoospermia, and they observed high levels of these two miRNAs in infertile patients compared to fertile individuals [133]. In conclusion, the miR-I9B and let-7 bis may represent good molecular markers for non-obstructive azoospermia cases with primary infertility or oligozoospermia. In according with these results, Dr. Tsatsanis *et al.* showed that the miR-155 serum was correlated with male fertility regardless of the systemic inflammation grade or androgenic alteration [134]. In this context, pollution or genotoxic, genetic and epigenetic stress can cause the spermiogenesis damage and this concern can be related to the susceptibility to chronic diseases in adulthood, but also and especially to the vulnerability to diseases of future generations (transgenerational effects) [135].

HUMAN SEMEN: ENVIRONMENTAL AND HEALTH MARKER

Epidemiological and experimental data demonstrate, how described above, that the male reproductive system is particularly and uniquely sensitive to a broad variety of environmental pollutants. Spermatogenesis unlike oogenesis from puberty onward is continuously and therefore more easily exposed to insults in his stages of continuous replication and so male germline accumulates mutations faster than female one [136, 137]. The oxidative stress is capable of damaging sperm cells more than egg. The sperm cells do not possess a significant antioxidant protection since the reduction of cytoplasmic space does not allow for an appropriate system of defensive enzymes and significant amounts of polyunsaturated fatty acids [138, 139]. Furthermore, in semen, in the same time, it is possible to estimate environmental contaminants and to evaluate *in vivo* effects on sperm cells, since they have features sensitive to environmental pollutants such as motility, morphology and the integrity of the DNA strand. Human semen with respect to other biological matrices (blood, urine, hair) is also a bio-accumulator. especially of heavy metals (our data in phase of submission) and of both exogenous and endogenous Volatile organic compounds VOCs [140]. Already, Rubes *et al.* conducted a study on differences of exposure of police officers who worked in the Centre of Prague (Czech Republic), and the authors found that sperm DNA fragmentation was significantly higher in winter (high exposure) in comparison in spring (low exposure) in samples of all men, including non-smokers [141]. Another study reported that the oxidative stress and apoptosis were induced in testis of Male Wistar rat treated with the 1,1,1-trichloro-2-2-bis(4-chlorophenyl) ethane (p,p'-DDT) (persistent organic pollutant). The authors found an increase in lipid peroxidation LPO level and hydrogen peroxide (H_2O_2) production and a decrease of metallothioneins levels, superoxide dismutase (SOD), catalase (CAT) activities. These results clearly suggested that DDT sub-acute treatment causes oxidative stress in rat testis leading to apoptosis [142]. Another study demonstrated the presence of high concentrations of heavy metals such as cadmium, chromium, lead, copper and zinc in the blood of fishermen living in in Abu Qir Bay, Egypt. This has been linked to significantly high levels of metallothionein in the fishermen's erythrocytes [143]. In the metropolitan area of Naples, several studies reported the association between low sperm motility and high environmental exposure such as emissions of traffic or heavy metals [144]. In addition, several studies within EcoFoodFertility project have demonstrated the role of human semen as an early and sensitive biomarker of environmental pollution. In fact significantly higher level of sperm DNA damage, measured by means two different techniques, was found in healthy male volunteers, no smokers, homogeneous for age and lifestyle living in the "Land of Fires" (High Environmental Impact Area, HEIA) as compared with that measured in volunteers living in same Region, Valle del Sele in province of Salerno (Low Environmental Empact Area, LEIA) [28, 145]. Human semen sensitivity to pollution-induced alteration of semen redox status was confirmed in another published study and conducted always in the same region of Italy [28]. In particular, it was shown that semen was more susceptible when compared with blood plasma to both pollution-associated alteration of redox status and Sperm Telomere Length (STL), while Leukocyte Telomere Length (LTL), was not significantly influenced by the environmental impact [146]. In another study of EcoFoodFertility was compared the seminal parameters (number, motility, morphology and index of spermatic DNA fragmentation) of 327 healthy males exposed both professionally (workers of the ILVA of Taranto), and residents in high environmental impact areas, Land of Fires in Campania Region and Taranto in Apulia Region, with those residents in two areas of low environmental impact areas, Alto Medio-Sele and Cilento in Campania Region and Palermo in Sicily. Where the atmospheric pollution measured through the PM10, PM2.5 and benzene levels was higher, there was a sperm DNA damage calculated by more than 30% in ILVA workers, and in the residents of the city of Taranto and the area of the Land of Fires (province of Naples and Caserta), compared to those of the other two control areas. This paper demonstrates that the first insult of a polluted environment goes directly to the spermatic DNA, and therefore, it turns out to be the seminal parameter more sensitive than the others (number, motility and morphology) and also, unfortunately, the one that carries the genetic information with the imaginable consequence [147]. Additionally, another Chinese study reported the relationship between air pollution and alterations in sperm morphology [148].

If human semen seems an earlier and sensitive source of biomarkers in comparison with blood to monitor high environmental pressure on human health [149], for this reason, a reliable environmental sentinel as shown in literature, can be considered the human semen, being an important health marker (Fig. **2**). In fact, the spermatogenesis cycle is many complex and susceptible to endogenous and exogenous stress, so it can represent a good indicator of the well-being state of the organism. Several studies reported the relationship between semen quality and state of health, by associating the semen quality with either chronic degenerative diseases, comorbidities and even mortality [81 - 88].

Among all reports, the Eisenberg studies linked semen quality with mortality rates [150, 151] since men with low sperm volume, concentration and sperm motility, had higher death rates than men with normal sperm parameters. In this prospective, fertility assessment may be an indicator of overall health and the attention on maximum fertility age (18-35 years) and it may be important for chronic diseases prevention. In addition to the potential preventive and predictive role of reproductive biomarkers in chronic adult degenerative diseases, the growing interest in the transgenerational effects, induced by pollution and lifestyles through epigenetic modifications on gametes, shifts the interest of prevention and therefore the interest towards reproductive biomarkers assumes a greater significance to safeguard the health of future generations. Moreover, if many results report that a healthy environment and lifestyle of the mother play a crucial role in offspring health, the utero window can represent a field of study of Developmental Origins of Health and Disease (DOHaD) [152, 153] since epigenetic modifications seem to be involved in mediating between environment impacts and early life health, and consequently, can be interesting to investigate the Paternal Origins of Health and Disease paradigm (POHaD) [154]. This term has been introduced on the basis of transgenerational epigenetic effect of contaminants through the paternal germ line on animal studies. Although there are few epidemiological studies on humans, the perspective of the systematic studies of reproductive biomarkers in environmental impact assessment and early and predictive health risk assessment can make a significant contribution in this field [155 - 160].

Fig. (2). Human semen: Environmental and Health marker.

CONCLUSION

The study of reproductive biomarkers such as seminal ones, that are extremely sensitive to environmental stress, early and even predictive of chronic degenerative diseases, can represent a keystone in assessing health risk for a revolution in the epidemiological field. In this perspective, the first signs of damage to organo-sentinel systems such as the reproductive system could allow to evaluate the development of different disease Moreover, such sensitive, early environmental and health markers, could help the policy makers to intervene promptly in areas with significant environmental criticalities. Furthermore, it would be possible to verify the effectiveness of the interventions by monitoring reproductive biomarkers. In this way, the implementation of effective measures would allow to safeguard the community and also its social and productive organization. Definitely, the double function of seminal marker (environmental and health sentinel), can represent more than an opportunity for public prevention policies in defense of current and future generations.

CONSENT FOR PUBLICATION

Not applicable.

CONFLICT OF INTEREST

The author declares that there is no conflict of interest in this chapter.

ACKNOWLEDGEMENTS

This chapter was performed within the "EcoFoodFertility" project (http://www.ecofoodfertility.it/)

REFERENCES

[1]   WHO Regional Office for Europe. Economic cost of the health impact of air pollution in Europe: Clean air, health and wealth. Copenhagen: WHO Regional Office for Europe 2015.

[2]   Risom L, Møller P, Loft S. Oxidative stress-induced DNA damage by particulate air pollution. Mutat Res 2005; 592(1-2): 119-37.
[http://dx.doi.org/10.1016/j.mrfmmm.2005.06.012]

[3]   Singh R, Sram RJ, Binkova B, *et al.* The relationship between biomarkers of oxidative DNA damage, polycyclic aromatic hydrocarbon DNA adducts, antioxidant status and genetic susceptibility following exposure to environmental air pollution in humans. Mutat Res 2007; 620(1-2): 83-92.
[http://dx.doi.org/10.1016/j.mrfmmm.2007.02.025] [PMID: 17445838]

[4]   Ojha A, Srivastava N. Redox imbalance in rat tissues exposed with organophosphate pesticides and therapeutic potential of antioxidant vitamins. Ecotoxicol E Mol Aspects Med 2011; 32(4-6): 234-46.
[PMID: 21864906]

[5]   Hybertson BM, Gao B, Bose SK, McCord JM. Oxidative stress in health and disease: the therapeutic potential of Nrf2 activation. Mol Aspects Med 2011; 32(4-6): 234-46.
[http://dx.doi.org/10.1016/j.mam.2011.10.006] [PMID: 22020111]

[6]   Williams MA, Rangasamy T, Bauer SM, *et al.* Georas SN Disruption of the transcription factor Nrf2 promotes pro-oxidative dendritic cells that stimulate Th2-like immune responsiveness upon activation by ambient particulate matter. J Immunol 2008; 1(7): 59-4545.

[7]   Faraonio R, Vergara P, Di Marzo D, *et al.* p53 suppresses the Nrf2-dependent transcription of antioxidant response genes. J Biol Chem 2006; 281(52): 39776-84.
[http://dx.doi.org/10.1074/jbc.M605707200] [PMID: 17077087]

[8]   Ruttkay-Nedecky B, Nejdl L, Gumulec J, *et al.* The role of metallothionein in oxidative stress. Int J Mol Sci 2013; 14(3): 6044-66.
[http://dx.doi.org/10.3390/ijms14036044] [PMID: 23502468]

[9]   Viarengo A, Burlando B, Dondero F, Marro A, Fabbri R. Aldo Viarengo Bruno Burlando Francesco Dondero Anna Marro Rita Fabbri. Metallothionein as a tool in biomonitoring programmes. Biomarkers 1999; 4(6): 455-66.
[http://dx.doi.org/10.1080/135475099230615] [PMID: 23902390]

[10]   Koedrith P, Seo YR. Advances in carcinogenic metal toxicity and potential molecular markers. Int J Mol Sci 2011; 12(12): 9576-95.
[http://dx.doi.org/10.3390/ijms12129576] [PMID: 22272150]

[11]   Haq F, Mahoney M, Koropatnick J. Signaling events for metallothionein induction. Mutat Res 2003; 533(1-2): 211-26.
[http://dx.doi.org/10.1016/j.mrfmmm.2003.07.014] [PMID: 14643422]

[12]   Carpenè E, Andreani G, Isani G. Metallothionein functions and structural characteristics. J Trace Elem Med Biol 2007; 21(1) (Suppl. 1): 35-9.
[http://dx.doi.org/10.1016/j.jtemb.2007.09.011] [PMID: 18039494]

[13]   Gomulkiewicz A, Podhorska-Okolow M, Szulc R, *et al.* Correlation between metallothionein (MT) expression and selected prognostic factors in ductal breast cancers. Folia Histochem Cytobiol 2010; 48(2): 242-8.
[http://dx.doi.org/10.2478/v10042-010-0011-5] [PMID: 20675281]

[14]   Jayasurya A, Bay BH, Yap WM, Tan NG, Tan BK. Proliferative potential in nasopharyngeal carcinoma: correlations with metallothionein expression and tissue zinc levels. Carcinogenesis 2000; 21(10): 1809-12.
[http://dx.doi.org/10.1093/carcin/21.10.1809] [PMID: 11023537]

[15]   Hengstler JG, Pilch H, Schmidt M, *et al.* Metallothionein expression in ovarian cancer in relation to histopathological parameters and molecular markers of prognosis. Int J Cancer 2001; 95(2): 121-7.

[http://dx.doi.org/10.1002/1097-0215(20010320)95:2<121::AID-IJC1021>3.0.CO;2-N]   [PMID: 11241323]

[16] Wülfing C, van Ahlen H, Eltze E, Piechota H, Hertle L, Schmid KW. Metallothionein in bladder cancer: correlation of overexpression with poor outcome after chemotherapy. World J Urol 2007; 25(2): 199-205.
[http://dx.doi.org/10.1007/s00345-006-0141-8] [PMID: 17253087]

[17] Weinlich G, Eisendle K, Hassler E, Baltaci M, Fritsch PO, Zelger B. Metallothionein - overexpression as a highly significant prognostic factor in melanoma: a prospective study on 1270 patients. Br J Cancer 2006; 94(6): 835-41.
[http://dx.doi.org/10.1038/sj.bjc.6603028] [PMID: 16508630]

[18] Si M, Lang J. The roles of metallothioneins in carcinogenesis. J Hematol Oncol. 2018 23;11(1):107.

[19] Ferrario C, Lavagni P, Gariboldi M, *et al.* Metallothionein 1G acts as an oncosupressor in papillary thyroid carcinoma. Lab Invest 2008; 88(5): 474-81.
[http://dx.doi.org/10.1038/labinvest.2008.17] [PMID: 18332874]

[20] Datta J, Majumder S, Kutay H, *et al.* Metallothionein expression is suppressed in primary human hepatocellular carcinomas and is mediated through inactivation of CCAAT/enhancer binding protein alpha by phosphatidylinositol 3-kinase signaling cascade. Cancer Res 2007; 67(6): 2736-46.
[http://dx.doi.org/10.1158/0008-5472.CAN-06-4433] [PMID: 17363595]

[21] Han YC, Zheng ZL, Zuo ZH, *et al.* Metallothionein 1 h tumour suppressor activity in prostate cancer is mediated by euchromatin methyltransferase 1. J Pathol 2013; 230(2): 184-93.
[http://dx.doi.org/10.1002/path.4169] [PMID: 23355073]

[22] Kerr J, Anderson C, Lippman SM. Physical activity, sedentary behaviour, diet, and cancer: an update and emerging new evidence Oncol 2017; 18(8): e457-e471.

[23] Steliarova-Foucher E, Fidler MM, Colombet M, *et al.* ACCIS contributors. Changing geographical patterns and trends in cancer incidence in children and adolescents in Europe, 1991-2010 (Automated Childhood Cancer Information System): a population-based study. Lancet Oncol 2018; 19(9): 1159-69.
[http://dx.doi.org/10.1016/S1470-2045(18)30423-6] [PMID: 30098952]

[24] Frazzoli C, Petrini C, Mantovani A. Sustainable development and next generation's health: a long-term perspective about the consequences of today's activities for food safety. Ann Ist Super Sanita 2009; 45(1): 65-75.
[PMID: 19567981]

[25] Pirastu R, Comba P, Conti S, *et al.* Sentieri –Epidemilogical Study of residents in National Priority Contaminated Sites: mortality, cancer incidence and hospital discharges. Epidemiol Prev 2014; 38(2) (Suppl. 1.).

[26] Pasetto R, Zengarini N, Caranci N, *et al.* [Environmental justice in the epidemiological surveillance system of residents in Italian National Priority Contaminated Sites (SENTIERI Project)]. Epidemiol Prev 2017; 41(2): 134-9.
[PMID: 28627155]

[27] Menchini-Fabris F, Rossi P, Palego P, Simi S, Turchi P. Declining sperm counts in Italy during the past 20 years. Andrologia 1996; 28(6): 304.
[http://dx.doi.org/10.1111/j.1439-0272.1996.tb02804.x] [PMID: 9021039]

[28] Bergamo P, Volpe MG, Lorenzetti S, *et al.* Human semen as an early, sensitive biomarker of highly polluted living environment in healthy men: A pilot biomonitoring study on trace elements in blood and semen and their relationship with sperm quality and RedOx status. Reprod Toxicol 2016; 66: 1-9.
[http://dx.doi.org/10.1016/j.reprotox.2016.07.018] [PMID: 27592743]

[29] Zhou N, Cui Z, Yang S, *et al.* Air pollution and decreased semen quality: a comparative study of Chongqing urban and rural areas. Environ Pollut 2014; 187: 145-52.

[http://dx.doi.org/10.1016/j.envpol.2013.12.030] [PMID: 24491300]

[30]   Nordkap L, Joensen UN, Blomberg Jensen M, Jørgensen N. Regional differences and temporal trends in male reproductive health disorders: semen quality may be a sensitive marker of environmental exposures. Mol Cell Endocrinol 2012; 355(2): 221-30.
[http://dx.doi.org/10.1016/j.mce.2011.05.048] [PMID: 22138051]

[31]   Mendiola J, Jørgensen N, Andersson AM, Stahlhut RW, Liu F, Swan SH. Reproductive parameters in young men living in Rochester, New York. Fertil Steril 2014; 101(4): 1064-71.
[http://dx.doi.org/10.1016/j.fertnstert.2014.01.007] [PMID: 24524829]

[32]   Hauser R, Sokol R. Science linking environmental contaminant exposures with fertility and reproductive health impacts in the adult male. Fertil Steril 2008; 89(2) (Suppl.): e59-65.
[http://dx.doi.org/10.1016/j.fertnstert.2007.12.033] [PMID: 18308066]

[33]   Triassi M, Alfano R, Illario M, Nardone A, Caporale O, Montuori P. Environmental pollution from illegal waste disposal and health effects: a review on the "triangle of death". Int J Environ Res Public Health 2015; 12(2): 1216-36.
[http://dx.doi.org/10.3390/ijerph120201216] [PMID: 25622140]

[34]   Senior K, Mazza A. Italian "Triangle of death" linked to waste crisis. Lancet Oncol 2004; 5(9): 525-7.
[http://dx.doi.org/10.1016/S1470-2045(04)01561-X] [PMID: 15384216]

[35]   Barba M, Mazza A, Guerriero C, *et al.* Wasting lives: the effects of toxic waste exposure on health. The case of Campania, Southern Italy. Cancer Biol Ther 2011; 12(2): 106-11.
[http://dx.doi.org/10.4161/cbt.12.2.16910] [PMID: 21734464]

[36]   Mazza A, Piscitelli P, Neglia C, Della Rosa G, Iannuzzi L. Illegal dumping of toxic waste and its effect on human health in campania, italy. Int J Environ Res Public Health 2015; 12(6): 6818-31.
[http://dx.doi.org/10.3390/ijerph120606818] [PMID: 26086704]

[37]   Crispo A, Barba M, Malvezzi M, *et al.* Cancer mortality trends between 1988 and 2009 in the metropolitan area of Naples and Caserta, Southern Italy: Results from a joinpoint regression analysis. Cancer Biol Ther 2013; 14(12): 1113-22.
[http://dx.doi.org/10.4161/cbt.26425] [PMID: 24025410]

[38]   Zaccaroni A, Corteggio A, Altamura G, *et al.* Elements levels in dogs from "triangle of death" and different areas of Campania region (Italy). Chemosphere 2014; 108: 62-9.
[http://dx.doi.org/10.1016/j.chemosphere.2014.03.041] [PMID: 24875913]

[39]   Montano L, Iannuzzi L, Rubes J, *et al.* EcoFoodFertility – Environmental and food impact assessment on male reproductive function. Andrology 2014; 2 (Suppl. 2): 69.

[40]   Critto A, Torresan S, Semenzin E, *et al.* Development of a site-specific ecological risk assessment for contaminated sites: part I. A multi-criteria based system for the selection of ecotoxicological tests and ecological observations. Sci Total Environ 2007; 379(1): 16-33.
[http://dx.doi.org/10.1016/j.scitotenv.2007.02.035] [PMID: 17439821]

[41]   Zinatloo-Ajabshir S, Mortazavi-Derazkola S, Salavati-Niasari M. Sonochemical synthesis, characterization and photodegradation of organic pollutant over Nd_2O_3 nanostructures prepared *via* a new simple route. Separ Purif Tech 2017; 178: 138-46.
[http://dx.doi.org/10.1016/j.seppur.2017.01.034]

[42]   Zinatloo-Ajabshir S, Mortazavi-Derazkola S, Salavati-Niasari M. Preparation, characterization and photocatalytic degradation of methyl violet pollutant of holmium oxide nanostructures prepared through a facile precipitation method. J Mol Liq 2017; 231: 306-13.
[http://dx.doi.org/10.1016/j.molliq.2017.02.002]

[43]   Zinatloo-Ajabshir S, Mortazavi-Derazkola S, Salavati-Niasari M. $Nd_2Sn_2O_7$ nanostructures: New facile Pechini preparation, characterization, and investigation of their photocatalytic degradation of methyl orange dye. Adv Powder Technol 2017; 28: 697-705.
[http://dx.doi.org/10.1016/j.apt.2016.11.017]

[44] Zinatloo-Ajabshir S, Mortazavi-Derazkola S, Salavati-Niasari M. New facile preparation of Ho2O3 nanostructured material with improved photocatalytic performance. J Mater Sci Mater Electron 2017; 28: 1914-24.
[http://dx.doi.org/10.1007/s10854-016-5744-2]

[45] Burger J. A framework for increasing sustainability and reducing risk to ecological resources through integration of remediation planning and implementation. Environ Res 2019; 172: 586-95.
[http://dx.doi.org/10.1016/j.envres.2019.02.036] [PMID: 30875512]

[46] Gochfeld, M., Burger, J., Powers, C., Kosson, D. Land use planning scenarios for contaminated land: comparing EPA, State, Federal and Tribal approaches. Waste Management Symposium, Waste Manage. Proc. Phoenix, Arizona, March 14-20, 2015.

[47] Cvitanovic C, McDonald J, Hobday AJ. From science to action: Principles for undertaking environmental research that enables knowledge exchange and evidence-based decision-making. J Environ Manage 2016; 183(Pt 3): 864-74.
[http://dx.doi.org/10.1016/j.jenvman.2016.09.038] [PMID: 27665124]

[48] Wcisło E, Bronder J, Bubak A, Rodríguez-Valdés E, Gallego JLR. Human health risk assessment in restoring safe and productive use of abandoned contaminated sites. Environ Int 2016; 94: 436-48.
[http://dx.doi.org/10.1016/j.envint.2016.05.028] [PMID: 27344373]

[49] Greenberg M, Lowrie K. A proposed model for community participation and risk communication for a DOE, led stewardship program. Fed Facil Environ J 2001; (Spring): 125-41.
[http://dx.doi.org/10.1002/ffej.3330120113]

[50] Burger J, Ed. Stakeholders and Scientists: Achieving Implementable Solutions to Energy and Environmental Issues. New York: Springer 2011.
[http://dx.doi.org/10.1007/978-1-4419-8813-3]

[51] Cundy AB, Bardos RP, Church A, et al. Developing principles of sustainability and stakeholder engagement for "gentle" remediation approaches: the European context. J Environ Manage 2013; 129: 283-91.
[http://dx.doi.org/10.1016/j.jenvman.2013.07.032] [PMID: 23973957]

[52] National Research Council (NRC). Understanding Risk: Informing Decisions in a Democratic Society. Washington, D.C.: National Academy Press 1996.

[53] Environmental Protection Agency (EPA). 2009.http://www.epa.gov/environmentaljustice

[54] Burger J. Environmental management: integrating ecological evaluation, remediation, restoration, natural resource damage assessment and long-term stewardship on contaminated lands. Sci Total Environ 2008; 400(1-3): 6-19.
[http://dx.doi.org/10.1016/j.scitotenv.2008.06.041] [PMID: 18687455]

[55] Burger J, Gochfeld M, Clarke J, Jeitner C, Pittfield T. Environmental assessment for sustainability and resiliency for ecological and human health. J Environ Stud (Northborough) 2015; 1(1): 1-8.
[PMID: 27468428]

[56] Cappuyns V. Inclusion of social indicators in decision support tools for the selection of sustainable site remediation options. J Environ Manage 2016; 184(Pt 1): 45-56.
[http://dx.doi.org/10.1016/j.jenvman.2016.07.035] [PMID: 27450992]

[57] Hull RN, Luoma SN, Bayne BA, et al. Opportunities and challenges of integrating ecological restoration into assessment and management of contaminated ecosystems. Integr Environ Assess Manag 2016; 12(2): 296-305.
[http://dx.doi.org/10.1002/ieam.1714] [PMID: 26419951]

[58] Burger J, Harris S, Harper B, Gochfeld M. Ecological information needs for environmental justice. Risk Anal 2010; 30(6): 893-905.
[http://dx.doi.org/10.1111/j.1539-6924.2010.01403.x] [PMID: 20409031]

[59]   Chan KMA, Satterield T, Goldstein J. Rethinking ecosystem services to better address and navigate cultural values. Ecol Econ 2012; 74: 8-13.
[http://dx.doi.org/10.1016/j.ecolecon.2011.11.011]

[60]   Bardos, P. Progress in sustainable remediation. Remediation, winter 2014, 23-32.
[http://dx.doi.org/10.1002/rem.21412]

[61]   Harclerode MA, Macbeth TW, Miller ME, Gurr CJ, Myers TS. Early decision framework for integrating sustainable risk management for complex remediation sites: Drivers, barriers, and performance metrics. J Environ Manage 2016; 184(Pt 1): 57-66.
[http://dx.doi.org/10.1016/j.jenvman.2016.07.087] [PMID: 27497675]

[62]   Lowrie K, Greenberg M, Simon D, Solitaire L, Killmer M, Mayer H. Remediation and stewardship: coexisting processes to protect health and the environment. Remediation 2003; 13: 91-104.
[http://dx.doi.org/10.1002/rem.10086]

[63]   Gunderson LH, Pritchard L Jr. Resilience and the behavior of large-scale systems SCOPE 60. Washington, D.C.: Island Press 2002.

[64]   Virapongse A, Brooks S, Metcalf EC, *et al.* A social-ecological systems approach for environmental management. J Environ Manage 2016; 178: 83-91.
[http://dx.doi.org/10.1016/j.jenvman.2016.02.028] [PMID: 27131638]

[65]   Gunderson LH, Reece C, Holling CS. Foundations of ecological resilience. Washington, D.C.: Island Press 2010.

[66]   Wagner AM, Larson DL, DalSoglio JA, *et al.* A framework for establishing restoration goals for contaminated ecosystems. Integr Environ Assess Manag 2016; 12(2): 264-72.
[http://dx.doi.org/10.1002/ieam.1709] [PMID: 26339869]

[67]   Wagner AM, Larson DL, DalSoglio JA, *et al.* A framework for establishing restoration goals for contaminated ecosystems. Integr Environ Assess Manag 2016; 12(2): 264-72.
[http://dx.doi.org/10.1002/ieam.1709] [PMID: 26339869]

[68]   Siddiqui IA, Jaleel A, Al'Kadri HM, Akram S, Tamimi W. Biomarkers of oxidative stress in women with pre-eclampsia. Biomarkers Med 2013; 7(2): 229-34.
[http://dx.doi.org/10.2217/bmm.12.109] [PMID: 23547818]

[69]   Neagu M, Albulescu R, Tanase C. Research highlights: highlights from the latest articles in biomarkers in medicine. Biomarkers Med 2013; 7: 201-4.
[http://dx.doi.org/10.2217/bmm.13.7]

[70]   Biomarkers Definitions Working Group. Biomarkers and surrogate endpoints: preferred definitions and conceptual framework. Clin Pharmacol Ther 2001; 69(3): 89-95.
[http://dx.doi.org/10.1067/mcp.2001.113989] [PMID: 11240971]

[71]   Vaidya VS, Ozer JS, Dieterle F, *et al.* Kidney injury molecule-1 outperforms traditional biomarkers of kidney injury in preclinical biomarker qualification studies. Nat Biotechnol 2010; 28(5): 478-85.
[http://dx.doi.org/10.1038/nbt.1623] [PMID: 20458318]

[72]   Carlsen E, Giwercman A, Keiding N, Skakkebaek NE. Evidence for decreasing quality of semen during past 50 years. BMJ 1992; 305(6854): 609-13.
[http://dx.doi.org/10.1136/bmj.305.6854.609] [PMID: 1393072]

[73]   Joffe M. Infertility and environmental pollutants. Br Med Bull 2003; 68: 47-70.
[http://dx.doi.org/10.1093/bmb/ldg025] [PMID: 14757709]

[74]   Rolland M, Le Moal J, Wagner V, Royère D, De Mouzon J. Decline in semen concentration and morphology in a sample of 26,609 men close to general population between 1989 and 2005 in France. Hum Reprod 2013; 28(2): 462-70.
[http://dx.doi.org/10.1093/humrep/des415] [PMID: 23213178]

[75]   Levine H, Jørgensen N, Martino-Andrade A, *et al.* Temporal trends in sperm count: a systematic

review and meta-regression analysis. Hum Reprod Update 2017; 23(6): 646-659.

[76]   Jensen TK, Jacobsen R, Christensen K, Nielsen NC, Bostofte E. Good semen quality and life expectancy: a cohort study of 43,277 men. Am J Epidemiol 2009; 170(5): 559-65.
[http://dx.doi.org/10.1093/aje/kwp168] [PMID: 19635736]

[77]   Merritt MA, De Pari M, Vitonis AF, Titus LJ, Cramer DW, Terry KL. Reproductive characteristics in relation to ovarian cancer risk by histologic pathways. Hum Reprod 2013; 28(5): 1406-17.
[http://dx.doi.org/10.1093/humrep/des466] [PMID: 23315066]

[78]   Baker JA, Buck GM, Vena JE, Moysich KB. Fertility patterns prior to testicular cancer diagnosis. Cancer Causes Control 2005; 16(3): 295-9.
[http://dx.doi.org/10.1007/s10552-004-4024-2] [PMID: 15947881]

[79]   Jørgensen N, Vierula M, Jacobsen R, *et al.* Recent adverse trends in semen quality and testis cancer incidence among Finnish men. Int J Androl 2011; 34(4 Pt 2): e37-48.
[http://dx.doi.org/10.1111/j.1365-2605.2010.01133.x] [PMID: 21366607]

[80]   Rives N, Perdrix A, Hennebicq S, *et al.* The semen quality of 1158 men with testicular cancer at the time of cryopreservation: results of the French National CECOS Network. J Androl 2012; 33(6): 1394-401.
[http://dx.doi.org/10.2164/jandrol.112.016592] [PMID: 22837112]

[81]   Eisenberg ML, Li S, Brooks JD, Cullen MR, Baker LC. Increased risk of cancer in infertile men: analysis of U.S. claims data. J Urol 2015; 193(5): 1596-601.
[http://dx.doi.org/10.1016/j.juro.2014.11.080] [PMID: 25463997]

[82]   Rogers MJ, Walsh TJ. Male infertility and risk of cancer. Semin Reprod Med 2017; 35(3): 298-303.
[http://dx.doi.org/10.1055/s-0037-1603583] [PMID: 28658714]

[83]   Hanson BM, Eisenberg ML, Hotaling JM. Male infertility: a biomarker of individual and familial cancer risk. Fertil Steril 2018; 109(1): 6-19.
[http://dx.doi.org/10.1016/j.fertnstert.2017.11.005] [PMID: 29307404]

[84]   Brinton LA. Fertility status and cancer. Semin Reprod Med 2017; 35(3): 291-7.
[http://dx.doi.org/10.1055/s-0037-1603098] [PMID: 28658713]

[85]   Choy JT, Eisenberg ML. Male infertility as a window to health. Fertil Steril 2018; 110(5): 810-4.
[http://dx.doi.org/10.1016/j.fertnstert.2018.08.015] [PMID: 30316415]

[86]   Barnhart KT. Introduction: Fertility as a window to health. Fertil Steril 2018; 110(5): 781-2.
[http://dx.doi.org/10.1016/j.fertnstert.2018.08.031] [PMID: 30316411]

[87]   Pisarska MD. Fertility status and overall health. Semin Reprod Med 2017; 35(3): 203-4.
[http://dx.doi.org/10.1055/s-0037-1603728] [PMID: 28658702]

[88]   Glazer CH, Bonde JP, Eisenberg ML, *et al.* Male infertility and risk of nonmalignant chronic diseases: A systematic review of the epidemiological evidence. Semin Reprod Med 2017; 35(3): 282-90.
[http://dx.doi.org/10.1055/s-0037-1603568] [PMID: 28658712]

[89]   Asklund C, Jørgensen N, Skakkebaek NE, Jensen TK. Increased frequency of reproductive health problems among fathers of boys with hypospadias. Hum Reprod 2007; 22(10): 2639-46.
[http://dx.doi.org/10.1093/humrep/dem217] [PMID: 17728352]

[90]   Jagai JS, Messer LC, Rappazzo KM, Gray CL, Grabich SC, Lobdell DT. County-level cumulative environmental quality associated with cancer incidence. Cancer 2017; 123(15): 2901-8.
[http://dx.doi.org/10.1002/cncr.30709] [PMID: 28480506]

[91]   Jensen TK, Jacobsen R, Christensen K, Nielsen NC, Bostofte E. Good semen quality and life expectancy: a cohort study of 43,277 men. Am J Epidemiol 2009; 170: 559-556.

[92]   Barazani Y, Katz BF, Nagler HM, Stember DS. Lifestyle, environment, and male reproductive health. Urol Clin North Am 2014; 41(1): 55-66.
[http://dx.doi.org/10.1016/j.ucl.2013.08.017] [PMID: 24286767]

[93]   Selevan SG, Borkovec L, Slott VL, *et al.* Semen quality and reproductive health of young Czech men exposed to seasonal air pollution. Environ Health Perspect 2000; 108(9): 887-94.
[http://dx.doi.org/10.1289/ehp.00108887] [PMID: 11017895]

[94]   Rubes J, Selevan SG, Evenson DP, *et al.* Episodic air pollution is associated with increased DNA fragmentation in human sperm without other changes in semen quality. Hum Reprod 2005; 20(10): 2776-83.
[http://dx.doi.org/10.1093/humrep/dei122] [PMID: 15980006]

[95]   Guven A, Kayikci A, Cam K, Arbak P, Balbay O, Cam M. Alterations in semen parameters of toll collectors working at motorways: does diesel exposure induce detrimental effects on semen? Andrologia 2008; 40(6): 346-51.
[http://dx.doi.org/10.1111/j.1439-0272.2008.00867.x] [PMID: 19032683]

[96]   Hammoud A, Carrell DT, Gibson M, Sanderson M, Parker-Jones K, Peterson CM. Decreased sperm motility is associated with air pollution in Salt Lake City. Fertil Steril 2010; 93(6): 1875-9.
[http://dx.doi.org/10.1016/j.fertnstert.2008.12.089] [PMID: 19217100]

[97]   Deng Z, Chen F, Zhang M, *et al.* Association between air pollution and sperm quality: A systematic review and meta-analysis. Environ Pollut 2016; 208(Pt B): 663-9.
[http://dx.doi.org/10.1016/j.envpol.2015.10.044] [PMID: 26552539]

[98]   Sonne SB, Kristensen DM, Novotny GW, *et al.* Testicular dysgenesis syndrome and the origin of carcinoma *in situ* testis. Int J Androl 2008; 31(2): 275-87.
[http://dx.doi.org/10.1111/j.1365-2605.2007.00855.x] [PMID: 18205797]

[99]   Skakkebaek NE, Berthelsen JG, Giwercman A, Müller J. Carcinoma-*in-situ* of the testis: possible origin from gonocytes and precursor of all types of germ cell tumours except spermatocytoma. Int J Androl 1987; 10(1): 19-28.
[http://dx.doi.org/10.1111/j.1365-2605.1987.tb00161.x] [PMID: 3034791]

[100]   Rapley EA, Crockford GP, Easton DF, Stratton MR, Bishop DT. International Testicular Cancer Linkage Consortium. Localisation of susceptibility genes for familial testicular germ cell tumour. APMIS 2003; 111(1): 128-33.
[http://dx.doi.org/10.1034/j.1600-0463.2003.11101171.x] [PMID: 12752252]

[101]   Hemminki K, Li X. Cancer risks in second-generation immigrants to Sweden. Int J Cancer 2002; 99(2): 229-37.
[http://dx.doi.org/10.1002/ijc.10323] [PMID: 11979438]

[102]   Bay K, Asklund C, Skakkebaek NE, Andersson AM. Testicular dysgenesis syndrome: possible role of endocrine disrupters. Best Pract Res Clin Endocrinol Metab 2006; 20(1): 77-90.
[http://dx.doi.org/10.1016/j.beem.2005.09.004] [PMID: 16522521]

[103]   Hauser R. The environment and male fertility: recent research on emerging chemicals and semen quality. Semin Reprod Med 2006; 24(3): 156-67.
[http://dx.doi.org/10.1055/s-2006-944422] [PMID: 16804814]

[104]   Jurewicz J, Hanke W, Radwan M, Bonde JP. Environmental factors and semen quality. Int J Occup Med Environ Health 2009; 22(4): 305-29.
[http://dx.doi.org/10.2478/v10001-009-0036-1] [PMID: 20053623]

[105]   Campion S, Aubrecht J, Boekelheide K, *et al.* The current status of biomarkers for predicting toxicity. Expert Opin Drug Metab Toxicol 2013; 9(11): 1391-408.
[http://dx.doi.org/10.1517/17425255.2013.827170] [PMID: 23961847]

[106]   Srám RJ, Binková B, Rössner P, Rubes J, Topinka J, Dejmek J. Adverse reproductive outcomes from exposure to environmental mutagens. Mutat Res 1999; 428(1-2): 203-15.
[http://dx.doi.org/10.1016/S1383-5742(99)00048-4] [PMID: 10517994]

[107]   Feng S, Jacobsen SE, Reik W. Epigenetic reprogramming in plant and animal development. Science 2010; 330(6004): 622-7.

[http://dx.doi.org/10.1126/science.1190614] [PMID: 21030646]

[108] Houshdaran S, Cortessis VK, Siegmund K, Yang A, Laird PW, Sokol RZ. Widespread epigenetic abnormalities suggest a broad DNA methylation erasure defect in abnormal human sperm. PLoS One 2007; 2(12):e1289.
[http://dx.doi.org/10.1371/journal.pone.0001289] [PMID: 18074014]

[109] Urdinguio RG, Bayón GF, Dmitrijeva M, *et al.* Aberrant DNA methylation patterns of spermatozoa in men with unexplained infertility. Hum Reprod 2015; 30(5): 1014-28.
[http://dx.doi.org/10.1093/humrep/dev053] [PMID: 25753583]

[110] Du Y, Li M, Chen J, *et al.* Promoter targeted bisulfite sequencing reveals DNA methylation profiles associated with low sperm motility in asthenozoospermia. Hum Reprod 2016; 31(1): 24-33.
[http://dx.doi.org/10.1093/humrep/dev283] [PMID: 26628640]

[111] Laurentino SS, Borgmann J, Gromoll J. On the origin of sperm epigenetic heterogeneity. Reproduction. 16. 2016; pii: REP-15-0436.

[112] Hammoud SS, Purwar J, Pflueger C, Cairns BR, Carrell DT. Alterations in sperm DNA methylation patterns at imprinted loci in two classes of infertility. Fertil Steril 2010; 94(5): 1728-33.
[http://dx.doi.org/10.1016/j.fertnstert.2009.09.010] [PMID: 19880108]

[113] Kobayashi H, Sato A, Otsu E, *et al.* Aberrant DNA methylation of imprinted loci in sperm from oligospermic patients. Hum Mol Genet 2007; 16(21): 2542-51.
[http://dx.doi.org/10.1093/hmg/ddm187] [PMID: 17636251]

[114] Marques CJ, Costa P, Vaz B, *et al.* Abnormal methylation of imprinted genes in human sperm is associated with oligozoospermia. Mol Hum Reprod 2008; 14(2): 67-74.
[http://dx.doi.org/10.1093/molehr/gam093] [PMID: 18178607]

[115] Khazamipour N, Noruzinia M, Fatehmanesh P, Keyhanee M, Pujol P. MTHFR promoter hypermethylation in testicular biopsies of patients with non-obstructive azoospermia: the role of epigenetics in male infertility. Hum Reprod 2009; 24(9): 2361-4.
[http://dx.doi.org/10.1093/humrep/dep194] [PMID: 19477879]

[116] Poplinski A, Tüttelmann F, Kanber D, Horsthemke B, Gromoll J. Idiopathic male infertility is strongly associated with aberrant methylation of MEST and IGF2/H19 ICR1. Int J Androl 2010; 33(4): 642-9.
[PMID: 19878521]

[117] Wu W, Shen O, Qin Y, *et al.* Idiopathic male infertility is strongly associated with aberrant promoter methylation of methylenetetrahydrofolate reductase (MTHFR). PLoS One 2010; 5(11): e13884.
[http://dx.doi.org/10.1371/journal.pone.0013884] [PMID: 21085488]

[118] Rajender S, Avery K, Agarwal A. Epigenetics, spermatogenesis and male infertility. Mutat Res 2011; 727(3): 62-71.
[http://dx.doi.org/10.1016/j.mrrev.2011.04.002] [PMID: 21540125]

[119] Hammoud SS, Nix DA, Hammoud AO, Gibson M, Cairns BR, Carrell DT. Genome-wide analysis identifies changes in histone retention and epigenetic modifications at developmental and imprinted gene loci in the sperm of infertile men. Hum Reprod 2011; 26(9): 2558-69.
[http://dx.doi.org/10.1093/humrep/der192] [PMID: 21685136]

[120] Paradowska AS, Miller D, Spiess AN, *et al.* Genome wide identification of promoter binding sites for H4K12ac in human sperm and its relevance for early embryonic development. Epigenetics 2012; 7(9): 1057-70.
[http://dx.doi.org/10.4161/epi.21556] [PMID: 22894908]

[121] Dada R, Kumar M, Jesudasan R, Fernández JL, Gosálvez J, Agarwal A. Epigenetics and its role in male infertility. J Assist Reprod Genet 2012; 29(3): 213-23.
[http://dx.doi.org/10.1007/s10815-012-9715-0] [PMID: 22290605]

[122] Rousseaux S, Caron C, Govin J, Lestrat C, Faure AK, Khochbin S. Establishment of male-specific epigenetic information. Gene 2005; 345(2): 139-53.

[http://dx.doi.org/10.1016/j.gene.2004.12.004] [PMID: 15716030]

[123] Weber M, Hellmann I, Stadler MB, *et al.* Distribution, silencing potential and evolutionary impact of promoter DNA methylation in the human genome. Nat Genet 2007; 39(4): 457-66.
[http://dx.doi.org/10.1038/ng1990] [PMID: 17334365]

[124] Kleene KC. Patterns, mechanisms, and functions of translation regulation in mammalian spermatogenic cells. Cytogenet Genome Res 2003; 103(3-4): 217-24.
[http://dx.doi.org/10.1159/000076807] [PMID: 15051942]

[125] Gatewood JM, Cook GR, Balhorn R, Bradbury EM, Schmid CW. Sequence-specific packaging of DNA in human sperm chromatin. Science 1987; 236(4804): 962-4.
[http://dx.doi.org/10.1126/science.3576213] [PMID: 3576213]

[126] Hayashi K, Chuva de Sousa Lopes SM, Kaneda M, *et al.* MicroRNA biogenesis is required for mouse primordial germ cell development and spermatogenesis. PLoS One 2008; 3(3):e1738.
[http://dx.doi.org/10.1371/journal.pone.0001738] [PMID: 18320056]

[127] Maatouk DM, Loveland KL, McManus MT, Moore K, Harfe BD. Dicer1 is required for differentiation of the mouse male germline. Biol Reprod 2008; 79(4): 696-703.
[http://dx.doi.org/10.1095/biolreprod.108.067827] [PMID: 18633141]

[128] Huszar JM, Payne CJ. MicroRNA 146 (Mir146) modulates spermatogonial differentiation by retinoic acid in mice. Biol Reprod 2013; 88(1): 15.
[http://dx.doi.org/10.1095/biolreprod.112.103747] [PMID: 23221399]

[129] Kotaja N. MicroRNAs and spermatogenesis. Fertil Steril 2014; 101(6): 1552-62.
[http://dx.doi.org/10.1016/j.fertnstert.2014.04.025] [PMID: 24882619]

[130] Khazaie Y, Nasr Esfahani MH. MicroRNA and male infertility: A potential for diagnosis. Int J Fertil Steril 2014; 8(2): 113-8.
[PMID: 25083174]

[131] Salas-Huetos A, Blanco J, Vidal F, *et al.* Spermatozoa from patients with seminal alterations exhibit a differential micro-ribonucleic acid profile. Fertil Steril 2015; 104(3): 591-601.
[http://dx.doi.org/10.1016/j.fertnstert.2015.06.015] [PMID: 26143365]

[132] Wang C, Yang C, Chen X, *et al.* Altered profile of seminal plasma microRNAs in the molecular diagnosis of male infertility. Clin Chem 2011; 57(12): 1722-31.
[http://dx.doi.org/10.1373/clinchem.2011.169714] [PMID: 21933900]

[133] Wu W, Hu Z, Qin Y, *et al.* Seminal plasma microRNAs: potential biomarkers for spermatogenesis status. Mol Hum Reprod 2012; 18(10): 489-97.
[http://dx.doi.org/10.1093/molehr/gas022] [PMID: 22675043]

[134] Tsatsanis C, Bobjer J, Rastkhani H, *et al.* Serum miR-155 as a potential biomarker of male fertility. Hum Reprod 2015; 30(4): 853-60.
[http://dx.doi.org/10.1093/humrep/dev031] [PMID: 25740880]

[135] Nilsson EE, Sadler-Riggleman I, Skinner MK, Riggleman M, Skinner K. Environmentally induced epigenetic transgenerational inheritance of disease. Environ Epigenet 2018; 4(2):dvy016.
[http://dx.doi.org/10.1093/eep/dvy016] [PMID: 30038800]

[136] Ségurel L, Wyman MJ, Przeworski M. Determinants of mutation rate variation in the human germline. Annu Rev Genomics Hum Genet 2014; 15: 47-70.
[http://dx.doi.org/10.1146/annurev-genom-031714-125740] [PMID: 25000986]

[137] Blumenstiel JP. Sperm competition can drive a male-biased mutation rate. J Theor Biol 2007; 249(3): 624-32.
[http://dx.doi.org/10.1016/j.jtbi.2007.08.023] [PMID: 17919661]

[138] Aitken RJ, Gibb Z, Baker MA, Drevet J, Gharagozloo P. Causes and consequences of oxidative stress in spermatozoa. Reprod Fertil Dev 2016; 28(1-2): 1-10.

[http://dx.doi.org/10.1071/RD15325] [PMID: 27062870]

[139] Dutta S, Majzoub A, Agarwal A. Oxidative stress and sperm function: A systematic review on evaluation and management. Arab J Urol 2019; 17(2): 87-97.
[http://dx.doi.org/10.1080/2090598X.2019.1599624] [PMID: 31285919]

[140] Montano L, Notari T, Longo V, *et al.* Human semen: Excellent bioaccumulator of volatile organic compounds (VOCs). Arch Ital Urol Androl 2019; 91(1): 15.

[141] Rubes J, Rybar R, Prinosilova P, *et al.* Genetic polymorphisms influence the susceptibility of men to sperm DNA damage associated with exposure to air pollution. Mutat Res 2010; 683(1-2): 9-15.
[http://dx.doi.org/10.1016/j.mrfmmm.2009.09.010] [PMID: 19800896]

[142] Marouani N, Hallegue D, Sakly M, Benkhalifa M, Ben Rhouma K, Tebourbi O. p,p'-DDT induces testicular oxidative stress-induced apoptosis in adult rats. Reprod Biol Endocrinol 2017; 15(1): 26-40.

[143] Saad AA, El-Sikaily A, Kassem H. Metallothionein and glutathione content as biomarkers of metal pollution in mussels and local fishermen in abu qir bay, egypt. J Health Pollut 2017; 6(12): 50-60.

[144] De Rosa M, Zarrilli S, Paesano L, *et al.* Traffic pollutants affect fertility in men. Hum Reprod 2003; 18(5): 1055-61.
[http://dx.doi.org/10.1093/humrep/deg226] [PMID: 12721184]

[145] Montano L, Notari T, Raimondo S, *et al.* Campania region group research EcoFoodFertility. Evaluation of environmental impact on sperm DNA integrity by sperm chromatin dispersion test and p53 ELISA. Preliminary data (ECOFOODFERTILITY project). Reprod Toxicol 2015; 56: 20.
[http://dx.doi.org/10.1016/j.reprotox.2015.07.044]

[146] Vecoli C, Montano L, Borghini A, *et al.* Effects of highly polluted environment on sperm telomere length: A pilot study. Int J Mol Sci 2017; 18(8): 1703.
[http://dx.doi.org/10.3390/ijms18081703] [PMID: 28777293]

[147] Bosco L, Notari T, Ruvolo G, *et al.* Sperm DNA fragmentation: An early and reliable marker of air pollution. Environ Toxicol Pharmacol 2018; 58: 243-9.
[http://dx.doi.org/10.1016/j.etap.2018.02.001] [PMID: 29448163]

[148] Lao XQ, Zhang Z, Lau AKH, *et al.* Exposure to ambient fine particulate matter and semen quality in Taiwan. Occup Environ Med 2018; 75(2): 148-54.
[http://dx.doi.org/10.1136/oemed-2017-104529] [PMID: 29133596]

[149] Montano L, Begamo P, Andreassi MG, Lorenzetti S. The role of human semen as an early and reliable tool of environmental impact assessment on human health. Full Chapter in Final Book Title & ISBN: Spermatozoa - Facts and Perspectives, " 978-1-78923-171-7. InTechOpen June 13th2018.
[http://dx.doi.org/10.5772/intechopen.73231]

[150] Eisenberg ML, Li S, Behr B, *et al.* Semen quality, infertility and mortality in the USA. Hum Reprod 2014; 29(7): 1567-74.
[http://dx.doi.org/10.1093/humrep/deu106] [PMID: 24838701]

[151] Eisenberg ML, Li S, Behr B, Pera RR, Cullen MR. Relationship between semen production and medical comorbidity. Fertil Steril 2015; 103(1): 66-71.
[http://dx.doi.org/10.1016/j.fertnstert.2014.10.017] [PMID: 25497466]

[152] Suzuki K. The developing world of DOHaD. J Dev Orig Health Dis 2018; 9(3): 266-9.
[http://dx.doi.org/10.1017/S2040174417000691] [PMID: 28870276]

[153] Bianco-Miotto T, Craig JM, Gasser YP, van Dijk SJ, Ozanne SE. Epigenetics and DOHaD: from basics to birth and beyond. J Dev Orig Health Dis 2017; 8(5): 513-9.
[http://dx.doi.org/10.1017/S2040174417000733] [PMID: 28889823]

[154] Soubry A. POHaD: why we should study future fathers. Environ Epigenet. 2018; 26;4(2): dvy007.

[155] Soubry A. Epigenetics as a driver of developmental origins of health and disease: did we forget the fathers? BioEssays 2018; 40(1)

[http://dx.doi.org/10.1002/bies.201700113] [PMID: 29168895]

[156] Curley JP, Mashoodh R, Champagne FA. Epigenetics and the origins of paternal effects. Horm Behav 2011; 59(3): 306-14.
[http://dx.doi.org/10.1016/j.yhbeh.2010.06.018] [PMID: 20620140]

[157] Soubry A, Hoyo C, Jirtle RL, Murphy SK. A paternal environmental legacy: evidence for epigenetic inheritance through the male germ line. BioEssays 2014; 36(4): 359-71.
[http://dx.doi.org/10.1002/bies.201300113] [PMID: 24431278]

[158] Braun K, Champagne FA. Paternal influences on offspring development: behavioural and epigenetic pathways. J Neuroendocrinol 2014; 26(10): 697-706.
[http://dx.doi.org/10.1111/jne.12174] [PMID: 25039356]

[159] Zhao ZH, Schatten H, Sun QY. Environmentally induced paternal epigenetic inheritance and its effects on offspring health. Reprod Devel Med 2017; 1(2): 89-99.
[http://dx.doi.org/10.4103/2096-2924.216862]

[160] Xavier MJ, Roman SD, Aitken RJ, Nixon B. Transgenerational inheritance: how impacts to the epigenetic and genetic information of parents affect offspring health. Hum Reprod Update 2019; 25(5): 518-40.
[http://dx.doi.org/10.1093/humupd/dmz017] [PMID: 31374565]

Hazardous Waste, Health Problems, and Personal Well-being: An International Perspective

James G. Linn[1,*], **Debra R. Wilson**[2], **Jorge Chuaqui**[3] and **Thabo T. Fako**[4]

[1] *Optimal Solutions in Healthcare and International Development, USA*

[2] *Austin Peay State University, USA*

[3] *University of Valparaiso, Chile*

[4] *University of Botswana, Botswana, Southern Africa*

Abstract: The negative impact of hazardous waste on public health is a growing global problem. Exposure to toxic waste is associated with numerous serious physical health problems, compromised intellectual development, mental illness and reduced personal well-being. This comprehensive analysis summarizes current epidemiological findings on the relationship of hazardous waste exposure to a wide range of physical illnesses and it also explores the pathways through which random exposure to toxic substances affects intellectual development in children and adults' mental health and perceived quality of life. The relationships are interpreted through the lens of sustainable societal development and environmental health. Regional differences and similarities of the physical and mental health consequences of toxic waste exposure are discussed using studies from the United States, Europe, Africa, and Latin America.

Keywords: Anxiety, Depression, Environmental health, Hazardous waste, Health problems, Intellectual development, Mental health, Personal well-being, Sustainable development .

INTRODUCTION

The negative impact of hazardous waste on public health is a growing global problem [1]. Toxic waste and solid waste are primary concerns of the member countries of the World Health Organization [2]. Inadequate, antiquated and unsanctioned methods of waste disposal, of solid, electronic and hazardous waste, are problems in all nations. This increasingly involves the transborder shipment of hazardous waste from industrialized to developing countries [3].

[*] Corresponding author James G. Linn: Development of International and Optimal Solutions in Healthcare, USA; Tel: 609-256-2522; E-mail: jlinn87844@aol.com

Gabriella Marfe & Carla Di Stefano (Eds.)

Data on hazardous waste sites which could potentially pollute soil and groundwater is systematically gathered in high-income countries. The United States Environmental Protection Agency (EPA) compiled a priority list of 1240 hazardous waste sites nationwide [4]. It was estimated by the EPA that there were approximately 41 million people residing within a 4-mile radius of one of the priority list sites. Information gathered by 33 European Countries, identified 342,000 waste sites that had contaminated surrounding soil and water there were 5.7 hazardous waste sites per 10,000 residents. [5].

Exposure to toxic waste is associated with numerous serious physical health problems, compromised intellectual development, mental illness, negative outcomes of pregnancy, and reduced personal well-being [6]. Fazio and associates (2017) completed a systematic review of 57 epidemiological studies of the level of health and health problems of communities located near poorly managed hazardous waste sites in high, middle- and low-income countries. There was adequate evidence of a connection between exposure to petroleum company waste and otolaryngological, respiratory, digestive, and dermatological acute symptoms.

Also, the researchers found limited evidence of a causal effect of toxic waste exposure and the development of hepatic, bladder, breast and testis carcinoma. Further, there was limited evidence of negative birth outcomes resulting from residential proximity to hazardous waste dumps, including premature birth, low birth weight and abnormalities of the musculoskeletal, urogenital and connective tissue systems [7]. Observed that brain development in early childhood could be adversely affected by chemical pollutants with harmful sensory, motor, and cognitive outcomes. They maintain that despite these initial findings, few large-scale studies exist to provide needed confirmatory evidence. Similarly, Edelstein [8] reported that although there is significant professional literature showing the psycho-social impacts of environmentally hazardous facilities, the results of these studies are not incorporated in environmental impact statements, nor reviewed in regulatory agencies for the issuance of hazardous facility permits.

DEFINITION OF TERMS

Hazardous Waste

This analysis uses a comprehensive definition of hazardous waste. We define it as solids, liquids, gases and sludges, which are harmful to humans and the environment. The sources of hazardous waste can be industry, agriculture, the community and/or households. It is a particular kind of waste because it cannot be easily disposed. Specifically, it includes petrochemical, pesticide, mining, manufacturing, electronic, medical, human and animal waste.

Furthermore, it is characterized by its ignitability, reactivity, corrosivity and/or toxicity. In developing this definition of hazardous waste, we have relied on information from the U.S. Environmental Protection Agency [9], the Basel Convention [10] and renowned authorities on hazardous waste [11].

Health Problems Associated with Hazardous Waste

This study focuses on health problems (including behavioral and mental health problems) that are associated with exposure to hazardous waste. We are focusing on health conditions specified by the U.S. Environmental Protection Agency and others reported in the scientific literature on the health impacts of hazardous waste. These health outcomes include cancer, behavioral abnormalities, mental distress, genetic mutations, physiological malfunctions (*e.g.* respiratory conditions, heart attacks, renal failure) and physical deformations [1,9,12].

Personal Well-being

Personal well-being is an individual's assessment of the quality of their life. A person has a high level of personal well-being if they have a sense that things are generally going well in their life [13]. For this analysis, we are equating personal well-being with subjective well-being, a concept on which there has been empirical research for over 30 years [14]. Subjective well-being is assumed to have a cognitive component measured by life satisfaction and an affective component measured by avowed happiness or positive and negative affect. An individual's subjective well-being is significantly impacted by health and mental health problems caused by toxins emanating from hazardous waste. We take a holistic approach to the conceptualization and measurement of subjective well-being. There are assumed to be several determinants at the individual, community, societal and environmental levels, which impact a person's subjective well-being [15].

Hazardous Waste Crises, Sustainable Economic Development, and Environmental Health

Sustainable Economic Development

Hazardous waste is a bi-product of industrialization, highly evolved technology, urbanization and modern social structure. Beginning in the post-war period, economists and scientists began to voice concerns to policymakers about the need to maintain ecosystem balance and to preserve essential resources in industrialized countries and societies in the process of industrialization [16-18]. Over the past 40 years, regions in the northern hemisphere (North America, Europe, Asia) have rapidly expanded industrial and post-industrial growth while those in the Southern Hemisphere (Africa, South America, the Indian Sub-Continent and South East Asia) have rapidly industrialized, or they have sectors which have industries which are important components of the globalized industrial economy. Policymakers in the United Nations and elsewhere during this period were very concerned with the need to balance economic growth/development with the environment [19]. Out of this discussion, the United Nations Commission on Environment and Development drafted the manuscript entitled "Our Common Future" in 1987, which is more commonly known as the Brundtland report, which contained the widely accepted definition of the concept of sustainable development [20].

Sustainable development is the development that meets the needs of the present without compromising the ability of future generations to meet their own needs. It contains within it two key concepts:

- The concept of 'needs', in particular, the essential needs of the world's poor, to which overriding priority should be given; and
- The idea of limitations imposed by the state of technology and social organization on the environment's ability to meet present and future needs.

Over the past 30 years, the concept of sustainable development has evolved with an emphasis on the socially inclusive and environmentally sustainable economic expansion [21]. This is concretely expressed in the sustainable economic development goals for the period 2016 to 2030, which were endorsed by the United Nations for all countries of the world. Implicit in them is the assumption that everyone, no matter which social class, locality or gender, will not be threatened by toxic, hazardous waste. The goals include: 1) no poverty, 2) zero hunger, 3) good health and well-being, 4) quality education, 5) gender equality, 6) clean water and sanitation, 7) affordable and clean energy, 8) decent work, and economic growth, 9) industry, innovation and infrastructure, 10) reducing inequalities, 11) sustainable cities and communities, 12) responsible consumption and production, 13) climate action, 14) life below water, 15) life on land, 16) peace, justice and strong institutions, 17) partnerships for the goals [22].

Environmental Health

The sustainable development goals, if attained, will fulfill the requirements of environmental health. The WHO (2016) [23] defines the scope of environmental health as "all of the physical, chemical and biological factors external to a person, and all of the related factors impacting behaviors. It encompasses the assessment and control of those environmental factors that can potentially affect health". More specifically, with regard to hazardous waste, environmental health practitioners and scientists deal with the direct harmful effects of chemicals, radiation and biological agents. Further, they are concerned with the direct and indirect effects on an individual's health and well-being of the structural, psychological, social and cultural environment. This includes housing, urbanization, use of land and transportation [24].

Hazardous Waste Crises & Radioactive Accidents in Industrialized and Developing Societies

Despite the agreement among most nations on the definition and goals of sustainable economic development, and the valuable roles of environmental health scientists and practitioners, there are continuing hazardous waste catastrophes and radioactive contamination accidents in industrialized and developing societies that create profound health and mental health problems which undermine personal well-being and continue unresolved for decades. Among the most prominent of these incidents are: The Chernobyl disaster, the Mariana and Brumadinho dam failures, the Kingston coal fly ash slurry spill, and the Ivory Coast toxic waste disposal.

In April of 1986, the Chernobyl nuclear reactor suffered an uncontrolled reaction condition that produced a fire and a plume of fissionable material for over a week which spread over areas in the USSR and Western Europe [25,26]. It is estimated that approximately four hundred times greater radioactive debris was produced by Chernobyl than the bombs dropped over Hiroshima and Nagasaki [27]. The actual number of deaths and injuries resulting from the Chernobyl accident remains controversial due to the lack of consistent data [28]. Immediately after the catastrophe, it was reported that 237 people were diagnosed with severe radiation sickness. Of these individuals, 31 died over the next several months [29]. The IAEA [30] concluded that 15 people succumbed to thyroid cancer in subsequent years and that there would be as many as 4000 additional cancer deaths over time in the areas most impacted by the radiation produced by the explosion and fire. Metter [31] also concluded that the radiation disaster had profound and lingering mental health effects on the people living in the most contaminated areas. They manifested classic symptoms of depression *i.e.*, hopelessness and helplessness and many of the youth became addicted to drugs to cope with their belief that they would ultimately die from the effects of the radiation poisoning.

The Mariana (2015) and Brumadinho (2019) catastrophic dam failures in Brazil are linked in the sense that both incidents included deaths, injuries and potential long-term serious health problems & negative environmental impacts. Further, both facilities were owned by the Vale company and the Brazilian federal government charged high company officials with criminal negligence in both cases [32, 33].

In November of 2015, the Fundao dam near the city of Mariana in the state of Minas Gerais, Brazil ruptured, sending 60 million cubic meters of iron waste into the Doce River. Nineteen people in the nearby community of Bento Rodriquez were killed [32]. The iron waste sludge continued to move downstream approximately 620 meters until it reached the Atlantic Ocean. The tailings pollution killed thousands of fish and destroyed almost 1500 hectares of forest located along the river. The long-term health and well-being of communities impacted by the iron tailings sludge is now in question because of the massive quantities of heavy metals-arsenic, lead and mercury left behind [22].

The Brumadinho dam failure in January of 2019 resulted in greater human casualties (237 dead and 33 missings) and also massive environmental degradation-pollution of 300 kilometers of rivers. Also, like the Mariana dam rupture, it left behind major amounts of heavy metals, which will be a health danger for decades to communities in the watershed [34].

For both the Mariana and Brumadinho dam disasters, the health problems related to hazardous waste (heavy metals) polluting the rivers and soils in the areas will become known in the future. Meanwhile, individuals living in the vicinity of the ruptured dams may experience PTSD from the catastrophes and fear at the prospect of chronic diseases that might emerge in the future.

There are several parallels among the Kingston coal fly ash slurry spill, which occurred in Tennessee in December of 2008 and the Mariana and Brumadinho dam rupture catastrophes. All three incidents involved the breach of containment structures for industrial waste, which reportedly were regularly inspected and assumed by the organizations which were responsible for them to be secure [35]. Ultimately, the incidents were linked to deaths and due to the toxic metals and other materials present in the sludge that escaped from the containment structures, many people were seriously sickened and/or presumed to be at risk for life-threatening illness in the future.

The Kingston coal fly, ash slurry spill, was the largest incident of its kind in the history of the United States [36]. It released over 1 billion gallons of coal ash slurry, which is a bi-product of coal combustion that contains a number of toxic materials including mercury, arsenic and benzene [37]. More than 300 acres of land were covered by the spill, which damaged homes in its path and polluted the Emery and Clinch rivers [38]. Although, initially, there were no reported injuries resulting from the release of the coal fly ash slurry, a decade later many hundreds of workers involved in the cleanup claimed serious illness due to their contact with the toxic coal fly ash waste and another 30 workers were alleged to have died as a result of their handling the dangerous dried ash slurry. A U.S. Federal judge upheld their claims.

While the Chernobyl reactor explosion and the Mariana, Brumadinho and Kingston, dam failures produced historic amounts of radiation and hazardous waste, the negative health impacts were mostly in the country where the disaster originated. However, this was not the case with a massive toxic waste dumping that occurred in the Ivory Coast in July of 2006 [39]. The Ivory coast incident was a situation where a business (Trafigura) in an industrialized country (Switzerland) used a relatively poor and politically unstable developing nation (the Ivory Coast) as a hazardous waste dump, which caused deaths and thousands of casualties there.

In August of 2006, the Proba Koala-a tanker chartered by the multinational company Trafigura-arrived at the port city Abidjan, Ivory Coast. The ship carried 500 tons of hazardous waste, which included the toxic materials sodium sulfide, phenols and sodium hydroxide [39]. The hazardous waste was transferred to a local business, Compagnie Tommy, which used subcontractors to drop it at various locations in & around Abidjan-waste deposits, public refuse heaps and in drainage areas along roads in the city [40]. The toxic gas emitted by the hazardous waste immediately caused headaches, vomiting and skin burns. Over the subsequent week, there were 17 certified deaths and an estimated 30, 000 injuries linked to the toxic materials dumped by the Proba Koala [41]. Ultimately, Trafigura had to pay the government of the Ivory Coast 100 million Euros for the cleanup of the toxic waste and also was fined another 30 million Euros by the Dutch government, which had originally refused off-loading of the toxic waste. Despite the huge fines paid to the government of the Ivory Coast and the imprisonment of local officials who participated in the hazardous waste dumping, the citizens of Abidjan experienced the physical and psychological trauma of the incident and a loss of trust in their institutions because they were not protected by them [42].

Hazardous Waste, Intellectual Development, Mental Illness and Subjective Well-being

Currently, sudden death, acute physical symptoms and serious physical health problems associated with hazardous waste and radiation exposure are well understood by physicians and biomedical scientists [43]. However, we are still exploring the pathways through which random and systematic exposure to toxic substances and radiation affects intellectual development in children and adults' mental health and perceived quality of life [7].

Despite the limited research on the fetus, children and adolescents, the

Environmental Protection Agency (2018) takes the position that chemical exposure can have serious adverse effects on the physical and intellectual development of them. The EPA maintains that the determinants that influence their vulnerability are their level of development and engagement with the environment. The fetus is at greatest risk because their developing brains and other organs may be severely damaged. Furthermore, children from one to six are developing rapidly and because of their high metabolism and degree of activity may ingest the substantial chemical, which has a great impact because their bodies are small. It is reported that chemicals absorbed into the body of a child can negatively impact normal cell and organ development. Adolescents are also susceptible because they continue to grow and are very active in their polluted environment.

Research on the effects of hazardous waste and radiation exposure on the mental health and subjective well-being of adults is mostly observational. Workers and community residents impacted by the Mariana and Brumadinho dam failures in Brazil were reported to suffer from PTSD associated with the sudden destructive floods of toxic sludge and were anxious about future chronic diseases that they might have acquired from the hazardous waste exposure [32,34]. Similarly, residents of communities affected by the fly ash "spill" in Kingston, Tennessee and recovery workers who handled the hazardous waste without protection were fearful of developing serious illness caused by the heavy metals contained in the sludge. Also, many residents of Abidjan, Ivory Coast were observed to be in a state of trauma after experiencing injuries resulting from the toxic waste dumping from the Proba Koala and depressed because neighbors died from poisoning [42]. In each of these cases, victims and community residents experienced a greatly decreased sense of personal well-being because they experienced trauma, were fearful of the future and had lost faith in institutions which they expected to protect them. Perhaps the most systematic data collection on the mental health effects of living near a hazardous waste facility was done by researchers who interviewed former residents of the love canal neighborhood. The love canal landfill was one of the most contaminated toxic waste sites in the United States. [44]. Researchers reported high levels of demoralization, depression, anxiety and alcohol consumption among adults who were relocated from the site [45].

Scholarly publication on the mental health & subjective well-being of persons affected by radiation contamination accidents began with the work of Metter [31], in the United States, and continued with discussions by [46] following the Fukushima nuclear disaster which resulted in the evacuation of 171, 000 individuals and resulted in an estimated 1600 deaths. The massive relocation combined with uncertainties regarding future health produced many cases of PTSD, anxiety and depression among persons who lived in the contamination zone. Furthermore, women carried a greater burden of these mental health problems and destroyed a sense of personal well-being. The gender difference in mental health problems and personal well-being is attributed to several factors: 1) women are concerned about their future fertility, 2) women are more concerned about the physical well-being of their children, 3) women have greater attachment to homes that they had to leave and 4) women feel greater stigma because they believe they are perceived to be contaminated with radiation [47].

The Way Forward to Reduce the Impact of Hazardous Waste on Physical Health, Mental Health, Intellectual Development and Personal Wellbeingg

To eliminate or ameliorate the negative effects of hazardous waste on individuals' physical health, mental health, intellectual development and personal quality of life we must improve and expand monitoring, community education, policy-making and prosecution of environmental legal violations. This systematic approach should be applied in all nations and governed by local and international institutions. This will require the increased resolve of leaders at all levels to acquire additional resources that will be applied toward these goals and the attainment of sustainable development.

CONCLUDING REMARKS

Such review has found that sudden death, acute physical symptoms, and serious physical health problems related to hazardous waste and radiation exposure are currently well understood by physicians and biomedical scientists. Epidemiological research on these issues has been carried out effectively and systematically for many decades. Of course, statements of causality with regard to the physical effects of hazardous waste and radiation exposure are limited by the research designs of the studies, which only make it possible to gather data on subjects after a catastrophic event has occurred. Nevertheless, in the analyses, comparisons can be made to health status data gathered on similar individuals in the community or those being treated in a clinic/hospital.

Unfortunately, far less is known about the effects of hazardous waste and radiation exposure, either resulting from a catastrophic incident or from residing near a hazardous waste dump or nuclear facility, on intellectual development in children, or on adults' mental health and subjective well-being. In the United States, the lack of systematic data gathered on these topics may in part, be due to the fact that they are not included in environmental impact statements nor reviewed by regulatory agencies that issue hazardous facility permits. However, epidemiological research on intellectual disabilities, *i.e.*, learning and behavioral issues in children, finds that they are substantial public health problems in most nations. Furthermore, as the recent Fukushima nuclear disaster has shown, the impact of a radiation incident and subsequent mass evacuation can produce high levels of PTSD, anxiety and depression, contributing to both morbidity and mortality.

Therefore, we recommend that social scientists, biomedical researchers and healthcare professionals should mobilize and petition policymakers to obtain resources to conduct large sample studies which will provide definitive answers regarding the pathways through which exposure to toxic substances and radiation affect intellectual development in children and adults' mental health and perceived quality of life. We also recommend that these studies be guided by the theories of Lazarus and Folkman [48] on stress and coping. This will be facilitated by reading an update on the application of their theories by Schneiderman and associates [49] and more recent articles made accessible by the U.S. Department of Health and Human Services.

Finally, in concert with the goals of sustainable development and the mission of environmental health, these researchers should treat all individuals and communities in their studies with dignity and respect and realize that in many cases, they will be interviewing people from marginalized and traditional places as well as mainstream communities.

CONSENT FOR PUBLICATION

Not applicable.

CONFLICT OF INTEREST

The authors confirm that the contents of this chapter have no conflict of interest.

ACKNOWLEDGEMENTS

Declare none.

REFERENCES

[1]   Fazzo L, Minichilli F, Santoro M, *et al.* Hazardous waste and health impact: a systematic review of the scientific literature. Environ Health 2017; 16(1): 107-15.
[http://dx.doi.org/10.1186/s12940-017-0311-8] [PMID: 29020961]

[2]   World Health Organization. 2015.https://www.who.int/phe/health_topics/en/

[3]   Landigan P, Wright R, Cordero J. The NIEHS superfund research program: twenty-years of translational research for the public. Environ Health 2015; 123: 909-18.

[4]   Pohl HR, Tarkowski S, Buczynska A, Fay M, De Rosa CT. Chemical exposures at hazardous waste sites: Experiences from the United States and Poland. Environ Toxicol Pharmacol 2008; 25(3): 283-91.
[http://dx.doi.org/10.1016/j.etap.2007.12.005] [PMID: 21783864]

[5]   Van Liedekerke M, Prokop G, Rabl-Berger S. Progress in the management of contaminated sites in Europe   2019.https://www.eea.europa.eu/data-and-maps/indicators/progress-in-manageme-t-of-contaminated-sites-3

[6]   Bergman A, Heindel J, Jobling S, Zoeller RT. State of the sciences of endocrine disrupting chemicals 2012.https://www.who.int/iris/bitstream/10665/78101/1/9789241505031_eng.pdf?ua=1

[7]   Koger SM, Schettler T, Weiss B. Environmental toxicants and developmental disabilities: a challenge for psychologists. Am Psychol 2005; 60(3): 243-55.
        [http://dx.doi.org/10.1037/0003-066X.60.3.243] [PMID: 15796678]

[8]   Edelstein MR. Weight and weightlessness administrative court efforts to weigh psycho-social impacts of proposed environmentally hazardous facilities. Impact Assess 1984; 3(3): 7-14.
        [http://dx.doi.org/10.1080/07349165.1984.9725536]

[9]   United States Environmental Protection Agency. 2019.https://www.epa.gov/

[10]   Basel Convention on the control of transboundary movements of hazardous wastes and their disposal secretariat methodological guide for the undertaking of national inventories of hazardous wastes within the framework of the Basel Convention series/sbc no: 99/009 (e) Gene 2000 May;

[11]   Orloff K, Falk H. An international perspective on hazardous waste practices. Int J Hyg Environ Health 2003; 206(4-5): 291-302.
        [http://dx.doi.org/10.1078/1438-4639-00225] [PMID: 12971684]

[12]   Vrijheid M. Health effects of residence near hazardous waste landfill sites: a review of epidemiologic literature. Environ Health Perspect 2000; 108 (Suppl. 1): 101-12.
        [PMID: 10698726]

[13]   Ryff CD, Keyes CLM. The structure of psychological well-being revisited. J Pers Soc Psychol 1995; 69(4): 719-27.
        [http://dx.doi.org/10.1037/0022-3514.69.4.719] [PMID: 7473027]

[14]   Tov W. Diener, E. Subjective well-being.Encyclopedia of cross-cultural psychology. Malden, MA: Wiley-Blackwell 2013; pp. 1239-45.
        [http://dx.doi.org/10.1002/9781118339893.wbeccp518]

[15]   Naci H, Ioannidis PA. Evaluation of wellness determinants and interventions by citizen scientists. JAMA 2015; 314 (2). p. 121. ISSN 0098-7484.

[16]   Carson R. Silent Spring. Boston: Houghton Mifflin Harcourt 1962.

[17]   Boulding KE. The economics of the coming Spaceship Earth. Environmental Quality in a Growing Economy: Essays from the Sixth RFF Forum. In: John Hopkins University Press; Baltimore 1966; pp. 3-14.

[18]   Meadows DH, Meadows DL. The limit of the Growth A Report for THE CLUB OF ROME'S Project on the Predicament of Mankind 1972.

[19]   International Union for Conservation. 1980.http://www.environmentandsociety.org/mml/ iucn-e--world-conservation-strategy-living-resource-conservation-sustainable-development

[20]   Report of the World Commission on Environment and Development: Our Common Future Transmitted to the General Assembly as an Annex to document A/42/427 - Development and International Co-operation: Environment https://sustainabledevelopment.un.org/content/documents/5987our-common-future.pdf

[21]   Sachs J. The Age of Sustainable Development. New York: Columbia University Press 2015.
        [http://dx.doi.org/10.7312/sach17314]

[22]   United Nations. Mine tailings storage: Safety is no accident Geneva April 2015.https://gridarendal.website.s3.amazonaws.com/production/documents/:s_document/370/original/RRAminewaste_flyer_screen.pdf?1509538685

[23]   World Health Organization. 2016.http://www.euro.who.int/__data/assets/pdf_file/0003/317226/Waste-human-health-Evidence-needs-mtg-report.pdf

[24]   Novice R. 1999.http://www.euro.who.int/__data/assets/pdf_file/0003/109875/E66792.pdf

[25]   McCall C. Chernobyl disaster 30 years on: lessons not learned. Lancet 2016; 387(10029): 1707-8. [http://dx.doi.org/10.1016/S0140-6736(16)30304-X] [PMID: 27116266]

[26]   Mulvey S. The Chernobyl nightmare revisited 2006.http://news.bbc.co.uk/2/hi/europe/4918742.stm

[27]   Fischer, D. History of the International Atomic Energy Agency: the first forty years / by David Fischer. Vienna: The Agency, 1997. p. ; 24 cm. "A Fortieth Anniversary Publication." ISBN 92–0–102397–9. (https://www-pub.iaea.org/mtcd/publications/pdf/pub1032_web.pdf)

[28]   Shkolnikov VM, McKee M, Vallin J, *et al.* Cancer mortality in Russia and Ukraine: validity, competing risks and cohort effects. Int J Epidemiol 1999; 28(1): 19-29. [http://dx.doi.org/10.1093/ije/28.1.19] [PMID: 10195659]

[29]   Hallenbeck W. Radiation Protection. Washington, DC: CRC Press 1994.

[30]   IAEA Annual Report for 2010.https://www.iaea.org/publications/reports/annual-report-2010

[31]   Mettler F. Chernobyl's Legacy. IAEA Bull 2011; 47(2): 34-43.

[32]   Willis A, Batista Y, Freiras G. Brazil Hunts for Mudslide Victims After BHP-Vale Dam Bursts. Bloomberg   2015.https://www.bloomberg.com/news/articles/2015-11-05/samarco-says-dam-in--razil-burst-teams-are-working-on-site

[33]   Phillips D. "https://www.theguardian.com/world/2019/feb/06/brazil-dam-collapse-workers-say-they-warned-owners" "That's going to burst: Brazilian dam workers say they warned of disaster". The Guardian. Retrieved Wednesday, February 6, 2019.

[34]   De Sousa M, Jeantet D. Brazilian environmental group tests water after dam collapse 2019.https://www.apnews.com/adc552694c7f45f2a164a2edd612f2c9

[35]   Flessner D, Sohn P. 2009.https://www.timesfreepress.com/news/news/story/2009/jan/05/tennessee-early-warnings-ash-pond-leaks/202428/ flapping/2015/aug/09/nagasaki-anniversary-radiation-nucl-ar-mental-health

[36]   Matheny   Jim.   2018.https://www.wbir.com/article/news/historic-disaster-10-years-after-t-e-ash-spill/51-3125fb4d-93bc-4dd8-9ce1-63a449fa6ff9

[37]   Dewan S. Waste Spills at Another TVA Power Plant New York Times 2009.https://www.nytimes.com/2009/01/10/us/10sludge.html

[38]   Paine   A,   Sledge   C.   Flood   of   sludge   breaks   TVA   dikes 2008.https://www.commondreams.org/news/2008/12/24/flood-sludge-breaks-tva-dike

[39]   Bernard   J,   Stroobants   J.   How   Abidjan   became   a   dump 2006.https://www.theguardian.com/world/2006/oct/20/outlook.development

[40]   Voice   of   America.   Ivory   coast   Government   releases   toxic   waste   findings 2010.https://www.voanews.com/a/a-13-2006-11-23-voa22/319097.html

[41]   Leigh D, Hirsch A. 2009.https://www.theguardian.com/environment/2009/may/13/trafigura-ivory-coast-documents-toxic-waste

[42]   Amnesty   International.   The   toxic   truth   about   a   company   called   Trafigura 2012.https://reliefweb.int/report/c%C3%B4te-divoire/toxic-truth-about-com-any-called-trafigura-ship-called-probo-koala-and-dumping

[43]   Centers for Disease Control and Prevention. Fourth National Report on Human Exposure to Environmental Chemicals 2015. https://www.cdc.gov/biomonitoring/pdf/FourthReport_Updated Tables_Feb2015.pdf

[44]   New York Dept. 1981.https://www.health.ny.gov/environmental/investigations/love_canal/

[45]   Gensburg LJ, Pantea C, Fitzgerald E, Stark A, Hwang SA, Kim N. Mortality among former Love Canal residents. Environ Health Perspect 2009; 117(2): 209-16.
[http://dx.doi.org/10.1289/ehp.11350] [PMID: 19270790]

[46]   Martin   B.   Nuclear   fallout: the   mental   health   consequences   of   radiation 2015.https://www.theguardian.com/science/brain-

[47]   Smith K, Matheson F, Moineddin R, *et al.* Gender differences in mental health service utilization among respondents reporting depression in a national health survey. Health 2013; 10(5): 1561-71.
[http://dx.doi.org/10.4236/health.2013.510212]

[48]   Biggs A, Brough P, Drummond S. Lazarus and Folkman. Psychological Stress and Coping Theory. In: Cooper CL, Campbell J, (Eds.). Quick First published: 18 February 2017.

[49]   Schneiderman N, Ironson G, Siegel SD. Stress and health: psychological, behavioral, and biological determinants. Annu Rev Clin Psychol 2005; 1: 607-28. [Review].
[http://dx.doi.org/10.1146/annurev.clinpsy.1.102803.144141] [PMID: 17716101]

CHAPTER 8

Zero Waste Management as Key Factor for Sustainable Development

Gabriella Marfe[1,*], **Carla Di Stefano**[2] and **Arturo Hermann**[3]

[1] *Department of Scienze e Tecnologie Ambientali, Biologiche e Farmaceutiche, University of Campania "Luigi Vanvitelli," via Vivaldi 43, Caserta 81100, Italy*

[2] *Department of Hematology, "Tor Vergata" University, Viale Oxford 81, 00133 Rome, Italy*

[3] *Italian National Institute of Statistics (ISTAT), Viale Liegi 13, Rome, Italy*

Abstract: The United Nation's Sustainable Development Goals (SDGs) represent an important framework to confront different issues such as economic growth, reducing harmful pollution, improving resource efficiency and waste management. Specifically, the SDGs cannot ignore the ever-growing problem of waste. The current economic system favors a 'take-make-and-dispose' model of production and consumption, which is no longer possible. For this reason, it is essential a change in waste management. Zero waste is a new concept to face waste management in today's society. It is a whole-system approach to redesign the resource life cycles so, in this way. it is possible to reuse all materials. Such strategy allows to create a low-carbon resource efficient, resilient and socially inclusive economy, to respect the diversity of ecosystems and to promote biodiversity, which is reflected in all seventeen of the SDGs.

Keywords: Prevention, Reuse, Recycle, Responsibility, Sustainable Development Goals (SDGs), Waste management, Zero Waste.

OBJECTIVE WITHIN THE GLOBAL INITIATIVES FOR SUSTAINABLE DEVELOPMENT

In analyzing the most relevant policies for zero waste, it is essential to locate them with the broader initiative for sustainable development carried out by the United Nations. These initiatives, although they appear at times distant from the real contexts, are relevant because they try to set out an inclusive and coordinated framework of policy action for sustainable development. Such coordination is vital for the effectiveness of policy action because economic and environmental

* **Corresponding author Gabriella Marfe:** Department of Scienze e Tecnologie Ambientali, Biologiche e Farmaceutiche, University of Campania "Luigi Vanvitelli," *via* Vivaldi 43, Caserta 81100, Italy; Tel: +39 0823 275104; Fax: +39 0823 274813; E-mail: gabmarfe@alice.it

problems are closely intertwined. And, of course, the issue of urban waste makes no exception.

THE UN AGENDA 2030 FOR SUSTAINABLE DEVELOPMENT

The Historical Background

The first step towards building a strategy of sustainable development was constituted by the Report "Our Common Future", also known as "Bruntland Report", published in 1987 by the World Commission on Environment and Development (WCED) [1].

The Report identifies a number of strategic goals and introduces a definition of sustainable development, largely employed afterwards, as "A development which meets the needs of the current generations without compromising the ability of future generations to meet their own needs".

This initiative, which perhaps went a bit unnoticed, had paved the way for the more influential "The United Nations Conference on Environment and Development" held in Rio de Janeiro in 1992. In this Conference, a precise political commitment was set up for the realization of the objectives of economic growth, social justice and environment protection.

The results of the Rio Conference, despite impressive progress in various areas, have not been sufficient to really take off ground a project of sustainable development. It had been then promoted the somewhat "intermediate" initiative of the "Millennium Development Goals" (MDGs), to be realized by 2015. It was also decided, in particular in the 2010 Millennium Development Goals Summit, to launch a "post-2015 Agenda" of sustainable development. In 2012, a second "Rio Conference" was organized, the so-called "Rio+20", in which was planned the report: "The Future We Want", the "Agenda 2030 for Sustainable Development" (from now on "Agenda 2030). Then, after an intense activity involving supranational and national institutions, the resolution was adopted by the UN General Assembly on 25 September 2015, the 2030 Agenda* for Sustainable Development" (from now on "Agenda 2030").

* Links https://www.un.org/sustainabledevelopment/development-agenda/

https://sustainabledevelopment.un.org/post2015/transformingourworld

MAIN FEATURES OF THE AGENDA 2030

The Agenda 2030 envisions a significant, and for some aspects epochal, shift towards a globally coordinated action for building an equitable and sustainable society. In order to better convey the "spirit" of the Agenda, we report some selected points of the Resolution.

"Our Vision"

"In these Goals and targets, we are setting out a supremely ambitious and transformational vision. We envisage a world free of poverty, hunger, disease and want, where all life can thrive. We envisage a world free of fear and violence. A world with universal literacy. A world with equitable and universal access to quality education at all levels, to health care and social protection, where physical, mental and social well-being are all assured. A world where we reaffirm our commitments regarding the human right to safe drinking water and sanitation and where there is improved hygiene and where food is sufficient, safe, affordable and nutritious. A world where human habitats are safe, resilient and sustainable and where there is universal access to affordable, reliable and sustainable energy.

We envisage a world of universal respect for human rights and human dignity, the rule of law, justice, equality and non-discrimination; of respect for race, ethnicity and cultural diversity; and of equal opportunity permitting the full realization of human potential and contributing to shared prosperity. A world that invests in its children and in which every child grows up free from violence and exploitation. A world in which every woman and girl enjoys full gender equality and all legal, social and economic barriers to their empowerment have been removed. A just, equitable, tolerant, open and socially inclusive world in which the needs of the most vulnerable are met."

"A Call for Action to Change our World"

"*Seventy years ago, an earlier generation of world leaders came together to create the United Nations. From the ashes of war and division they fashioned this Organization and the values of peace, dialogue and international cooperation which underpin it. The supreme embodiment of those values is the Charter of the United Nations*"

Today we are also taking a decision of great historical significance.

We resolve to build a better future for all people, including the millions who have been denied the chance to lead decent, dignified and rewarding lives and to achieve their full human potential. We can be the first generation to succeed in ending poverty; just as we may be the last to have a chance of saving the planet. The world will be a better place in 2030 if we succeed in our objectives.

What we are announcing today – an Agenda for global action for the next 15 years – is a charter for people and planet in the twenty-first century. Children and young women and men are critical agents of change and will find in the new Goals a platform to channel their infinite capacities for activism into the creation of a better world.

" We the people" are the celebrated opening words of the Charter of the United Nations. It is "we the peoples" who are embarking today on the road to 2030. Our journey will involve Governments as well as parliaments, the United Nations system and other international institutions, local authorities, indigenous peoples, civil society, business and the private sector, the scientific and academic community – and all people. Millions have already engaged with, and will own, this Agenda. It is an Agenda of the people, by the people and for the people – and this, we believe, will ensure its success.

The future of humanity and of our planet lies in our hands. It lies also in the hands of today's younger generation who will pass the torch to future generations.

We have mapped the road to sustainable development; it will be for all of us to ensure that the journey is successful and its gains irreversible. (from the Resolution adopted by the UN General Assembly on 25 September 2015).

THE SUSTAINABLE DEVELOPMENT GOALS (SDGS)

With regard to the definition of the Agenda 2030 objectives, these have been discussed and elaborated in various initiatives involving many countries, experts and stakeholders. The final document, named *Open Working Group Proposal for Sustainable Development*, identifies 17 general goals (SDGs) and 169 more specific targets. For their monitoring, the 48th UN Statistical Commission (UNSC), in March 2017, approved a global list of 244 SDG indicators, of which 1/3 is available at present

https://unstats.un.org/sdgs/indicators/indicators-list/

The 17 SDGs are the following (Fig. **1**):

Goal 1 — End poverty in all its forms everywhere.

Goal 2 — End hunger, achieve food security and improved nutrition and promote sustainable agriculture.

Goal 3 — Ensure healthy lives and promote wellbeing for all at all ages.

Goal 4 — Ensure inclusive and equitable quality education and promote lifelong learning opportunities for all.

Goal 5 — Achieve gender equality and empower all women and girls.

Goal 6 —Ensure availability and sustainable management of water and sanitation for all.

Goal 7 — Ensure access to affordable, reliable, sustainable and modern energy for all.

Goal 8 — Promote sustained, inclusive and sustainable economic growth, full and productive employment and decent work for all.

Goal 9 — Build resilient infrastructure, promote inclusive and sustainable industrialization and foster innovation.

Goal 10 — Reduce inequality within and among countries.

Goal 11 — Make cities and human settlements inclusive, safe, resilient and sustainable.

Goal 12 — Ensure sustainable consumption and production patterns.

Goal 13 — Take urgent action to combat climate change and its impacts.

Goal 14 — Conserve and sustainably use the oceans, seas and marine resources for sustainable development.

Goal 15 — Protect, restore and promote sustainable use of terrestrial ecosystems, sustainably manage forests, combat desertification, and halt and reverse land degradation and halt biodiversity loss.

Goal 16 — Promote peaceful and inclusive societies for sustainable development, provide access to justice for all and build effective, accountable and inclusive institutions at all levels.

Goal 17 — Strengthen the means of implementation and revitalize the global partnership for sustainable development.

As emerges from the previous description, the Agenda 2030 constitutes a major shift towards a more comprehensive view of sustainable development.

Such perspective clearly acknowledges the structural linkages between economic imbalances (such as unemployment and huge income disparities) and environmental decay. As these aspects reinforce each other in a vicious circle, the central implication for policy action is that no enduring improvement of economic condition can be realized without ensuring the environmental sustainability.

As an apparent weak aspect, we can note that the Agenda 2030 (and the Paris Agreement as well) provides that its application be left to the responsibility of each country. In this respect, the role of UN is mainly one of promotion, general coordination, exhortation and monitoring. This is fine, but it can prove to be insufficient to sustain brand new course for issues that, although nationally and locally rooted, also comprise a paramount supranational dimension. This calls for a better policy coordination both at **(i)** the horizontal level between the various policies for instance, macroeconomic, research and innovation, social, environmental; and at **(ii)** the vertical level, between various institutional dimensions for instance, supranational, national, subnational, local.

In the present situation, where resurgent xenophobia and nationalism make it arduous to broaden the role of the UN, the most proximate solution is to encourage as much as possible the coordination procedures already provided in the Agenda.

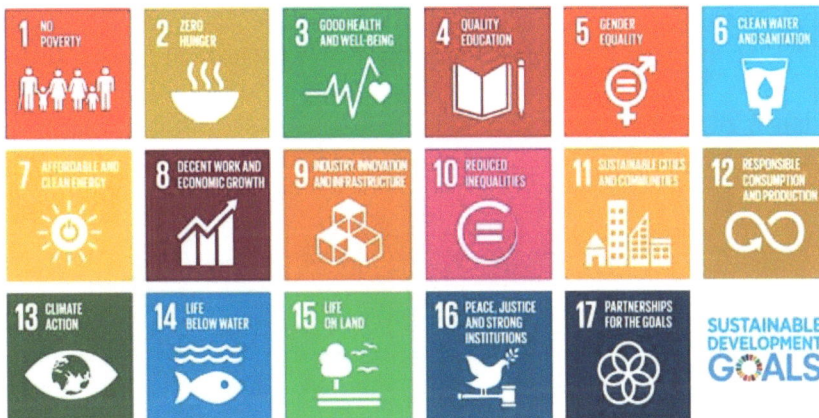

Fig. (1). The seventeen Sustainable Development Goals (SDGs) of the United Nations' Sustainable Development Agenda.

THE LINKS WITH ZERO WASTE INITIATIVES

What are the links of the Agenda 2030 and its SDGs with zero waste initiatives? Although the zero waste objective, it is not explicitly mentioned in the SDGs perhaps because at that time such problem was not yet so urgent it is obvious that it is implicitly contained in some of them (in particular 11 and 12). In this regard, considering the role of policy coordination underlined before, in our analysis of the main initiatives for zero waste, we will pay particular attention to their links with the measures of implementation of the Agenda 2030 at the national and supranational level.

WASTE MANAGEMENT IN LIGHT OF THE SDGS

The sustainable development concept intends "to seek equitable development between the current generation in the present and those in the future", as defined in the Brundtland Report [1]. Then, the sustainability concept implies that the needs of the present generation should not compromise the ability of future generations to meet theirs. Such perspective includes: 1) to protect the present and future generations' needs; 2) to avoid consumption of natural resources for human well-being; 3) to be aware of the natural extraction impacts [2]. This sustainable development has three main pillars: economic, environmental, and social [3]. Regarding these issues, in two different studies, El-Haggar [4] and Tudor *et al.* [5] point out that waste management must be effective, comprehensive and sustainable in order to create a self-sufficient community. As Farrelly *et al.* and Gupta *et al.* [6, 7] argue, "improving sustainable waste management practices to simultaneously improve livelihood opportunities, healthcare, education, and trade". The SDGs have set out some goals and targets that are related to waste. Now, virtually all SDGs, cannot be met unless waste management that is addressed as a priority. However, the specific goals for waste management are Goal 11 and 12.

Goal 11 "Make cities and human settlements inclusive, safe, resilient and sustainable".

It includes a target which focuses on reducing the adverse per capita environmental impact of cities by paying special attention to air quality and municipal and other waste management (target 11.6). By 2030, 5 billion people will live within cities and it is crucial to plane proper management practices to face the challenges brought by urbanization. These challenges, such as the safe removal and management of solid waste within cities, can be overcome in order to improve resource use and to decrease pollution and poverty. Furthermore, the energy efficiency of cities can be improved by waste management strategy, so that they become more sustainable in the long-term.

Goal 12 "Responsible consumption and production", and target goal 12.5 "By 2030, substantially reduce waste generation through prevention, reduction, recycling and reuse" [8]. The reuse offers a large potential for waste prevention and, in addition, the products can be used again for the same purpose for which they were conceived. (Directive 2008/98/EC of the European Parliament and of the Council of 19 November 2008 on waste and repealing certain Directives. Brussels, Belgium: European Parliament and the Council of the European Union). An economic study reported that producers prefer the production of new items in order to obtain a higher profit. Cooper and Timothy (2015) [9] evaluated that lower taxes for labour and higher taxes for energy and raw materials could help to promote re-use.

These goal includes targets focused on environmentally sound management of all waste through prevention, reduction and reuse as a reference to the waste hierarchy (targets 12.4 and 12.5) and reduction of food waste (12.3).

In a paper, Rodic and Wilson [10] believe that proper waste management can play an important role for achieving SDGs. Therefore, all SDGs should be incorporated into national waste strategies (Fig. **2**).

Goal 1 — End poverty in all its forms everywhere.

SDG 1 aims to eliminate poverty. It becomes possible by creating new jobs for people. Proper waste management can result in valuable materials to reuse. In this way it is possible to save money while potentially creating new jobs and business opportunities. Today, 1% of the global urban population make their living from recovering, recyclable materials from waste.

Goal 2 — End hunger, achieve food security and improved nutrition and promote sustainable agriculture.

SDG 2 aims to achieve food security through food waste reduction. According to WWF (World Wildlife Fund 2016), "a dietary shift in high-income countries – through consuming less animal protein – and reducing waste along the food chain could contribute significantly to producing enough food within the boundaries of one planet" [11].

Goal 3 — Ensure healthy lives and promote wellbeing for all at all ages.

SDG 3 includes target focused on reduction of the number of deaths and illnesses from hazardous chemicals and air, water and soil pollution and contamination (target 3.9). Waste is poisoning the and when people have no safe waste management, they can dump waste in the open or burn it. In this case, the health impacts of open burning are catastrophic.

Goal 4 — Ensure inclusive and equitable quality education and promote lifelong learning opportunities for all.

SDG 4 aims to make learning opportunities accessible to all. It also examines the quality of education, which plays a large role in sustainable development and poverty alleviation.

The increasing amount of waste can lead to significant global ecological risks as such as the shortage of natural resources, climate changes, mass pollution of water, air and soil. This issue can be solved through education that makes human beings knowledgeable to environmental problems. For this reason, a good waste management acknowledgment has to be covered with educational plans.

Goal 5— Achieve gender equality and empower all women and girls.

SDG 5 aims to achieve gender equality. In this case, women can benefit from accurate waste management, through earning opportunities and protecting their families from sickness caused by open dumping and burning.

Goal 6 —Ensure availability and sustainable management of water and sanitation for all.

SDG 6 includes target focused on improvement water quality by reducing pollution, eliminating dumping and minimizing release of hazardous chemicals and materials, halving the proportion of untreated wastewater and substantially increasing recycling and safe reuse globally (target 6.3).

This goal aims to facilitate access to fresh water, in sufficient quantity and quality, for all individuals. New investments in adequate infrastructure of waste management practices and sanitation facilities are necessary to allow universal access to safe drinking water by 2030. Furthermore, it will be essential to protect water-related ecosystems such as forests, mountains, wetlands and rivers to prevent water scarcity.

Goal 7 — Ensure access to affordable, reliable, sustainable and modern energy for all.

SDG 7 aims to increase substantially the share of renewable energy in the global energy mix, and to double the global rate of improvement in energy efficiency. Today it is possible to convert trash into energy—so-called waste-to-energy (WTE)- through different technologies. For example, waste from the manufacture of food products can be used to fed animals, and inedible remains can be converted into biogas and clean renewable energy.

Goal 8 — Promote sustained, inclusive and sustainable economic growth, full and productive employment and decent work for all.

SDG 8 promotes decent work for all women and men, including young people and persons with disabilities. It means to protect labour rights and improve working environments for all workers, including waste workers. Moreover, it focuses on improvement of global resource efficiency in consumption and production in association with economic growth without further environmental deterioration, in accordance with the 10-year framework of programmes on sustainable consumption and production, with developed countries taking the lead (target 8.4).

Goal 9 — Build resilient infrastructure, promote inclusive and sustainable industrialization and foster innovation.

SDG 9 aims to sustainable industrialization and promoting innovation across company operations, businesses can contribute to development efforts in the regions in which they operate through upgrading local infrastructure, investing in resilient energy and communications technologies, and making these technologies available to all people. It includes focused on upgrade infrastructure and retrofit industries to make them sustainable, with increased resource-use efficiency and greater adoption of clean and environmentally sound technologies and industrial processes, with all countries taking action in accordance with their respective capabilities (target 9.4). Thus, it is also crucial to invest in sustainable waste management. Today, this sector employs 20 million of people globally and it is currently the hotbed of innovations in recycling and recovery through new infrastructures. Informal waste management plays a critical role in collection and bringing recyclables back into the value chain. In this regard, many workers cannot have minimum wages, social security benefits, and personal protective equipment and other safety.

Goal 10 — Reduce inequality within and among countries.

SDG 10 includes goal focused on equal opportunity and inequalities reduction of outcome, including by eliminating discriminatory laws, policies and practices and promoting appropriate legislation, policies and action in this regard (target 10.3). Improper waste management means that all citizens of the world in live in an unsafe and unhealthy environment Today, many developing countries are unable to resolve the waste management in appropriate manner to resolve social inequality.

Goal 13 — Take urgent action to combat climate change and its impacts.

SDG 13 aims to reduce the factors that induce climate change. For instance, the impact of methane and CO_2 derived from poorly managed waste could be responsible for the increase for up to a tenth of greenhouse gases that contribute to the climate change.

Goal 14 - Conserve and sustainably use the oceans, seas and marine resources for sustainable development.

SGD 14 aims to face the threats of marine and nutrient pollution, resource depletion and climate change. These threats place further pressure on environmental systems, like biodiversity and natural infrastructure. Specifically, it includes goal focused on prevention and significantly reduction marine pollution of all kinds, in particular from land-based activities, including marine debris and nutrient pollution (target 14.1). Innovative solutions are essential to promote ocean sustainability. Today, about 13 million metric tons of plastic ended up in the oceans causing the plastic debris ingestion, starvation and drowning of marine animals Furthermore, the microplastics have been found in foodstuffs and drinking water, also in mountain soils in Switzerland [12]. Furthermore, several studies suggest that plastic particles can enter the body through ingestion or inhalation. In this scenario, the accumulation of microplastics could cause localized toxicity by inducing or exacerbating an immune response. Indeed, in human samples of both non-neoplastic and malignant lung tissues derived from biopsies, were detected the presence of cellulosic and plastic micro-fibers [13]. In another study, conducted in arctic ice core samples, the authors detected the presence of microplastics, which were derived from ocean currents from the Pacific garbage patch, or from local pollution from shipping and fishing [14].

Goal 15 — Protect, restore and promote sustainable use of terrestrial ecosystems, sustainably manage forests, combat desertification, and halt and reverse land degradation and halt biodiversity loss.

SGD 15 aims to reduce or reverse the effects of human activity on forests and other parts of the environment and to provide a more viable ecological platform for sustainable development. Specifically, it includes goal focused on the conservation, restoration and sustainable use of terrestrial and inland freshwater ecosystems and their services, in particular forests, wetlands, mountains and drylands, in line with obligations under international agreements target (target 15.1). In this context, groundwater, streams and rivers are poisoned by toxic chemicals that are contained of hazardous dumped waste. Life on land can only be possible when waste is safely managed.

Goal 16 — Promote peaceful and inclusive societies for sustainable development, provide access to justice for all and build effective, accountable and inclusive institutions at all levels.

SGD 16 aims to promote peaceful and inclusive societies and to build effective and accountable to enable the effective, efficient and transparent mobilization and use of resources. Specifically, it includes goal focused on reduction corruption and bribery in all their forms (target 16.5).

For instance, the rewards of proper waste management far outweigh the cost in comparison with the illegal waste management. For all communities, governments must urgently invest in a proper waste management to live in healthy places. Even the poor choose to pay for waste management (or participate in it) when they see its benefits. Furthermore Producer Responsibility schemes and fiscal transparency- would allow pay their right taxes in order to keep the planet clean (https://www.iswa.org/fileadmin/galleries/Publications/ISWA-Reports/GWMO_summary_web.pdf)

Goal — 17 Strengthen the means of implementation and revitalize the global partnership for sustainable development.

SGD 17 aims to implement and revitalize the global partnership for sustainable development. Specifically, it includes goal focused on promotion of the development, transfer, dissemination and diffusion of environmentally sound technologies to developing countries in concessional and preferential terms, as mutually agreed (Technology -- target 17.7). In this context, it could be important the diffusion of clean technologies for sustainable waste management to developing countries from developed countries. Furthermore, another goals intends to enhance the global partnership for sustainable development, complemented by multi-stakeholder partnerships that mobilize and share knowledge, expertise, technology and financial resources, to support the achievement of the sustainable development goals in all countries, in particular developing countries (Multi-stakeholder partnerships --target 17.16).

Sustainable waste management will be improved by sharing knowledge, promoting the creation and transfer of environmentally sound technologies. In this regard, it will be important to collaborate and work in partnership.

In summary, zero waste strategy is reflected in all seventeen of the SDGs since it aims to create a low-carbon, resource, efficient, sustainable and socially inclusive economy. Above all. such strategy allows to respect both ecosystems diversity and biodiversity.

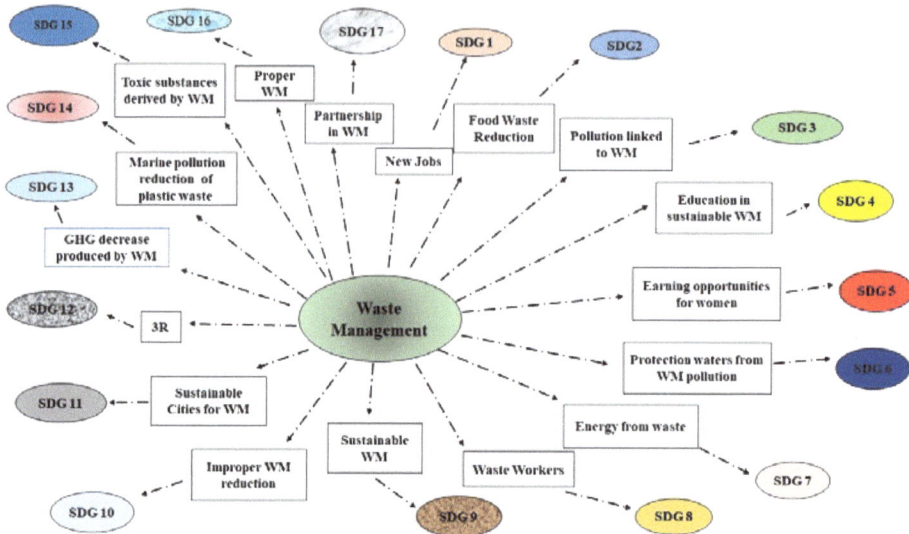

Fig. (2). SDGs in association with waste management.

SUSTAINABLE WASTE MANAGEMENT

"Until the environmental movement emerged in the 1960s, most waste was disposed of with little or no control: to land, as open burning; to air, by burning or evaporation of volatile compounds; or to water, by discharging solids and liquids to surface, groundwater or the ocean" [15].

As the above quote indicates, the sustainable waste management is a great challenge throughout the world. Many developing countries are still using landfills, incineration, and open dumps [16]. Furthermore, improper waste management will result in many risks to human health as well as to the degradation of the environment.

THE CONCEPT OF SUSTAINABLE WASTE MANAGEMENT

The sustainability of the environment as well as the engagement of society should adhere to waste hierarchy, that is a guideline for sustainable waste management. As argued by Wilson [17], "the waste hierarchy can also be seen as a 'historical' first step towards a current move away from the 'end-of-pipe' concept of 'waste management', towards the more integrated concept of 'resource management'. The principles of 'reduce, reuse, and recycle' (3Rs) have become a fundemantal step in sustainable waste management efforts for individual practice. The 3Rs placed in the second and third step of a waste hierarchy [18]. In a study, Gertsakis and Lewis [19] point out that reusing materials rather than creating new products avoid new waste production. Furthermore, they specify that energy, necessary to recycle waste into new products, should be placed at the third stage of the waste hierarchy. To date, another important concept to achieve sustainability is prevention. Such task is very difficult to realize since firstly it needs to change excessive consumerism culture [20]. Indeed, Davies *et al.* [21] report that the "prevention of waste involves a change in lifestyle to decrease waste production. Furthermore, industrial production should be able to redesign packaging to create more durable and nontoxic products. One of the techniques to handle organic waste is composting, which can be included in the fourth step of waste recovery (Fig. **3**). At the bottom of the waste hierarchy is landfill Open dumping has negative impacts as Townsend *et al.* [22] note, in terms of human health and environmental pollution including potential uncontrolled, leachate, air and water pollution. Nevertheless, many developing countries still use open dumping as their final disposal method because of the lack of infrastructures and economic resources [23]. Solid waste is not only municipal, but there are various fractions of hazardous waste derived from health-care facilities (hospital waste). Such waste contains about 10–25% of hazardous waste that poses a variety of environmental and health risks. In developing countries, open dumping is the most common method of hospital waste (HW) disposal, being low cost option. Such dangerous disposal is easily accessible both to waste pickers and other people who may be subject to infections of pathogenic microorganisms through indirect contact, inhalation, or ingestion. Moreover, the burning of this waste diminishes its volume but at the same time, produces toxic compounds like PCDD/F (polychlorinated dibenzo-p-dioxin and polychlorinated dibenzofuran) among other pollutants. For example, in the West Bank (Palestine), a study reports that 82.2% of HW is disposed of in unsafe dumpsites [24], while in Ibadan, Nigeria, more than 60% of HW handlers do not separate the HW and municipal solid waste (MSW) during collection and handling stages [25]. In Dhaka, Bangladesh, HW is collected by waste pickers to search for recyclables and reusable items (syringes, blades, knives, saline bags, plastic materials and metals) [26]. Therefore, in developing countries, waste management is becoming an environmental and social issue.

ZERO WASTE AS THE IDEAL CONCEPT IN SUSTAINABLE WASTE MANAGEMENT

Today, the application of advanced technology does not guarantee waste reduction [27]. For this reason, it is essential to find innovative and sustainable ways of planning urban life by governments and non-governmental institutions.

Zero waste is an alternative paradigm based on fully recovering of every kind of waste through the use of appropriate materials and design for durability. The global zero waste system has emerged in response to mounting waste-related problems such as pollution, health impacts, and the loss of land to new landfills. There are many zero waste initiatives around the world, primarily at the local government level that deals with waste management whose results often unsatisfactory.

THE CONCEPT OF ZERO WASTE

An integrated waste management approach attempts to solve waste problem by considering the entire life cycle of a product and determining the best method in order to extract as much useful material while saving energy, water, and other resources. For this reason, it is necessary a complete rethinking of "waste", which calls for:

WASTE to become WEALTH

REFUSE to become RESOURCE

TRASH to become CASH

The usual definition of zero waste was proposed by the Zero Waste International Alliance in 2004 [28]: *Zero Waste is a goal that is ethical, economical, efficient and visionary, to guide people in changing their lifestyles and practices to emulate sustainable natural cycles, where all discarded materials are designed to become resources for others to use. Zero Waste means designing and managing products and processes to systematically avoid and eliminate the volume and toxicity of waste and materials, conserve and recover all resources, and not burn or bury them. Implementing Zero Waste will eliminate all discharges to land, water or air that are a threat to planetary, human, animal or plant health.*

The term 'zero waste', used for the first time by Palmer [29] in 1973, indicated a method to recover resources from chemical waste. It is important to invest in zero waste infrastructure to avoid the building of new huge landfills or incinerators. Such strategy is a real integrated Waste Management System in which each material that is recovered from the waste stream is seen as a resource and managed according to its unique properties [28 - 30]. At the beginning, zero waste concept was introduced to reduce the amount of solid waste to dispose at landfill or send to incineration. For example, in Sweden the manufacturing and selling of mercury thermometers were banned in 1993-1994, being a source of mercury (toxic substance) in waste. In 1994, new rules were introduced both in MSW and the responsibility of producer to increase the amount of recycling and reuse of materials. Today, such system should allow to turn this issue into an opportunity by working to recycle waste and help to formalize the waste management sector. Furthermore, up to 90% of resources could be recovered from waste, if managed by trained personnel. About 60% of total waste could be transformed into compost, which can replenish soil nutrients, or biogas. Remaining wastes, if they are properly processed, can be recycled and transformed into new products.

In zero waste the 4 R's (Reduce, Reuse, Recycle and Responsibility) are famous, but the Responsibility can play a fundamental role in zero waste development in the world. It needs individual responsibility, community responsibility, industrial responsibility, professional responsibility and political responsibility.

In the field of industries, it could be important to pursue:

1) Design for Sustainability,

2) Clean Production

3) Extended Producer Responsibility.

In the first stage, the products and their constituent materials should be recyclable [31]. In the second stage, the use of toxic elements (such as lead, cadmium, mercury *etc.*) and compounds (such as chlorine, bromine and fluorine) should be eliminated in manufacture [32]. In the last stage, manufacturers and retailers could take back their products for reusing and recycling. Different companies are embarking on zero waste strategies, saving several millions money [33, 34]. Community Responsibility is based on source separation and door-to-door collection systems [35, 36]. Specifically, it is important to collect a clean organic fraction to obtain a compost, that can be used by farmers to replenish their soils of depleted nutrients. For example, in San Francisco, the kitchen and other organic waste (including the large fraction derived from restaurants and hotels) is transferred to a large composting plant located close to farmlands [35, 36]. In Italy, door-to-door collection systems are developed to generate clean organics in different cities [36]. Another example is Zurich (Switzerland), where a number of households (ranging from 3-200) can participate in a program called "community composting" and handle a simple compost system [37]. In zero waste strategy, the recyclable materials are sent to Material Recovery Facilities (or MRFs). Here, paper, cardboard, glass, metals and plastic are separated and prepared to be employed as secondary materials to manufacture new products. Some of these plants can handle a single stream of mixed recyclables (paper products) (*e.g.* Perth, Australia) or two streams (bottles, cans *etc.*) (*e.g.* Edmonton, Alberta, Canada). At the end, the reduction of residuals is a very important step in this system and for this reason, there are many Local and National Waste Reduction Initiatives. In this regard, especially the packaging of products needs to increase by introducing new rules. In Ireland, the government introduced a 15-cent tax on each plastic shopping bag used in shopping malls. In this manner, the use of these bags was reduced by 92% and the remaining 8% put over 12 million euros into funds for other recycling initiatives [38]. In Australia, 80 towns have banned the use of plastic shopping bags completely [39] as well as in San Francisco [40]. In some supermarkets, in Italy, food, drinks, shampoos, liquid washes and detergents are sold from dispensers or in reusable containers [41]. A considerable proportion of the total waste streams is represented by plastic waste that must be abated in many countries (20 kg/capita/year in Asia, 100 kg/capita/year in Western Europe and North America, 16 kg/capita/year in Africa [42, 43]. This kind of waste contains a wide range of additives that have toxic effects on human health (EDCs; *e.g.* bisphenol A or diethylhexylphthalate (DEHP); persistent organic pollutants (POPs), polybrominated diphenyl ethers (PBDEs), hexabromobiphenyls (HBB), hexabromocyclododecane (HBCD), short-chain chlorinated paraffins (SCCPs) and the fluorinated tensides like perfluorooctanoic acid (PFOA). For instance, in Vietnam, it was launched the Zero Plastic Waste City project that was initiated as a collaboration between The Grameen Creative Lab and The Alliance to End Plastic Waste. Such program aims to increase the waste collection rates of currently unconsidered waste types and increase the amount of waste being reused for new purposes. The implementation of this project will occur in specific locations close to rivers or the sea, and where it is present a high volume of solid waste leakage. In particular, the project will be primarily implemented in small and medium-sized urban areas [44]. Furthermore, in different countries, disposable bottles are replaced with reusable glass bottles [44]. In this regard, the zero waste strategy provides for the establishment of Reuse, Repair and Retraining Centers. A very good example of a reuse operation is given by a not-for-profit agency "Recycling North" in Burlington, Vermont. This agency grosses nearly $1 million a year and employs 27 people. They provide both free goods and job training for poor people. For instance, they train people to repair goods in five different categories: large appliances; small appliances; electrical goods; electronics and computers [45, 46]. Furthermore, Recycling North receives tax-deductible donations. These are few examples of these centers, but they are growing up around the world [46]. Another essential stage in the zero waste strategy is the deconstruction: such task consists of reusing and repairing old buildings. For example, recovered materials (such as doors and windows) can be reused to make new items like furniture [47, 48]. In Halifax, Nova Scotia, the Renovators Resource Center sells furniture made from recovered materials [49]. Another niche industry reuses and recycles the materials yielded by the renovation of hotels and office buildings (*e.g.* Sonrise Recycling in California) [50]. The final stage of normal waste management provides to send the residual to a landfill or an incinerator, but the zero waste strategy has established another step called Pay by bag systems to divert and decrease the residuals production. Specifically, the residual fractions can be picked up by applying an extra charge. It occurs in different ways: the residuals can be weighed, or it can be placed on the curb in a bag with purchased specific labels or put in special plastic bags that are purchased. This one simple fiscal step has led to significant reductions in many municipalities [51, 52]. The last stage of the Zero Waste strategy is the Residue Facility: In a "Zero Waste – Or Darn Near" world, a realistic goal is that the residue is less than 10%. If the majority of the organics and hazardous wastes have been treated with the appropriate technologies, the residue will result inert and may be disposed of in a "dry tomb" landfill with little potential risk (http://www.zerowastekauai.net/uploads/3/4/8/8/34884832/zero_waste_resource_guide_net.pdf)

For example, in Nova Scotia, the bags containing the residuals are sent in a building close to a landfill. In this building, the well-trained personnel open the bags and put the contents onto conveyor belts, and pull out bulky items, more recyclables and more toxics. Then, the untouched dirty organic fraction is biologically stabilized by a second composting operation, or through an anaerobic digestion system in other facilities. In this way, it is possible to control all degradation process before its underground disposal [53 - 55]. Naturally, these landfills must be built with more sophisticated engineering methods to contain both gaseous and liquid effluents (leachate). Furthermore, the Residual Screening facility should be located close to local university or technical college: 1) to improve the capture rate of reusables, recyclables and clean compostable in the door to door collection systems; 2) to develop alternative compounds for manufacturing of products (no toxic substances); 3) to offer new industrial designs on packaging and products. Several studies have divided the most important indicators of the zero waste strategy into seven main domains, namely geo-administrative, socio-cultural, management, environment, economy, organization, government and policy as shown in Fig. (**3**) [56]. Conversely, other studies have suggested to take action directly on the consumption of resources, individuals' consumption behaviour, product design based on cradle-to-cradle principles (eco-effective product and system design), maximum waste diversion from landfill and optimum resource recovery. For example, waste diversion rate (determined by calculating the amount of waste diverted from landfill) could be a performance assessment indicator for different waste management systems [57]. But such indicator presents limited forecasting capacity [58] for socio-economic and environmental issues [59, 60]. For this reason, two authors suggested an alternative performance assessment tool: zero waste index (ZWI). It allows to measure virgin material substitution by waste management systems. Globally, such index can measure the virgin material offset potentiality and the possible depletion of natural resources. In this context, it is possible to compare different waste management systems in multiple cities by considering the potential demand for virgin materials, energy, carbon pollution and water. A high ZWI value means greater substitution of virgin materials, energy and water saving. Zaman and Lehmann, in their study, have calculated the ZWI through one equation that measures the potentiality of virgin materials to be offset by zero waste management systems [59].

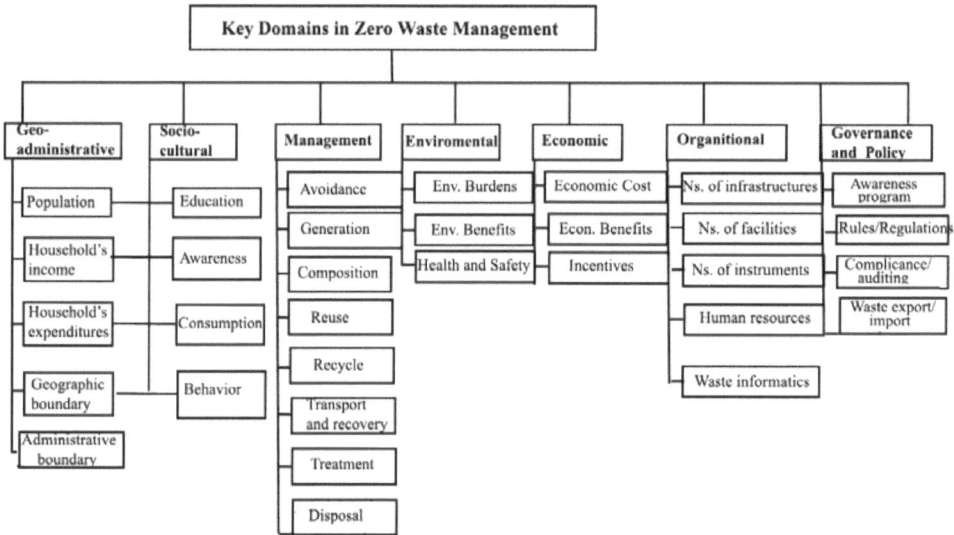

Fig. (3). Important Domais in Zero Waste Strategy (Nizar M, Munir E, Munawar E, Matseh I, Waller V. Applying Zero Waste Management Concept in a City of Indonesia: A Literature Review International Journal of Engineering & Technology, 2018, 7 (4), 6072-6077).

$$\text{Zero Waste Index} = \frac{\sum \text{Potential of waste managed by the city} + \text{substitution for the systems}}{\text{Total of waste generated in the city}}$$

Such index represents the value of the material that can potentially replace the virgin material inputs. The substitution values for material, energy, water and greenhouse gases (GHG) emissions were calculated from the life cycle database of different life cycle assessment tools and database sources. Furthermore, the advancement of technology used in the material recovery process is positively associated with the amount of both materials and resources substituted. In this regard, the substitution value depends on the utilized materials and different waste management systems. However, such index can be limited because there are not sufficient studies on quantitative measurement of waste prevention by behaviour change (Table **1**). Furthermore, the two authors found that San Francisco has a higher zero waste index (0,51) than Adelaide (0,23) and Stockholm (0,17) [59]. Another study calculated zero waste index value in the Faculty of Civil Engineering and Planning (FCEP) Islamic University of Indonesia (UII) after applying zero waste program since September 2016. The authors reported that the value of zero waste index at FCEP UII was 0,26. Such value indicated that this campus reused 134,19 kg waste of total 516,37 kg waste that has been produced. In this way, waste management system in the campus potentially substituted the energy demand of 1139,64 megajoules (MJ). Before this project, the garbage of campus was disposed of landfill and produced 132,91 kg of CO_2 for week. After this project, the garbage of campus produced 55.9 kg CO_2 for week because paper, plastics, and organic waste were reused. Specifically, paper, plastic and organic waste lessened the emission of greenhouse gas (GHG) by 17.69 kg, 22.68 kg, 36.64 kg CO_2 per week respectively. In conclusion, this study showed that FCEP UII should improve in virgin materials substitution and resource recovery [61]. In this scenario, to better assess zero waste management in a city, it is essential to understand the local context and the global market. The city government will have to provide a plan where it will specify the necessary investments to have the Advanced Waste Treatment (AWT) facilities. However, the amount of generated waste and the composition of urban solid waste have to be associated with the income of the community, as the higher the income, the more people consume and throw away. Today, the World Bank estimates that the total quantity of waste generated in low-income countries is expected to increase by more than three times by 2050 (http://datatopics.worldbank.org/what-awaste/trends_in_solid_ waste_management.html). About 23% and 6% of world's waste is generated by the East Asia and Pacific region and Middle East and North Africa region, respectively. By 2050, It has been estimated that waste generation of Sub-Saharan Africa, South Asia, and the Middle East and North Africa is more than triple, double, and double, respectively. In these regions, the waste management is unsafe and unhealthy because more than half of waste is currently openly dumped. This system may pose negative implications for the environment, health, and prosperity, thus requiring urgent action. The waste collection plays a critical role in managing waste: in low-income countries, 48% and 26% of waste is collected in cities and outside of urban areas, respectively. Across regions, about 44% of waste is collected in Sub-Saharan Africa, while in Europe and Central Asia and North America, about 90% of waste is collected (http://datatopics.worldbank.org/what-a-waste/trends_in_solid_waste_manage-ment.html). High-income countries generate about 32% of waste derived from food and green waste, while 51%, derived from dry waste, can be recycled, including plastic, paper, cardboard, metal, and glass. Middle- and low-income countries generate 53% and 57% food and green waste, respectively. Furthermore, in low-income countries, only 20% of the waste stream can be recycled. Finally, all regions generate about 50% or more organic waste, on average, while Europe and Central Asia and North America can generate higher portions of dry waste (http://datatopics.worldbank.org/ what-a-waste/trends_in_solid_waste_management.html). In this situation, the integration of community participation supported by strong regulations, and a strategy at the local context will be determining factors to realize zero waste system.

Table 1. Possible substitution of resources considering the zero waste index (adapted by Zaman AU. Measuring waste management performance using the "Zero Waste Index": The case of Adelaide, Australia', Journal of Cleaner Production. 2014b; (66), pp. 407–409).

Waste System	Material Type	Efficiency in Tonnes Virgin Material Substitutions	Efficiency in Tonnes Energy Substitutions	Greenhouse Gas Emissions Reduction (Tonnes/CO_2)	Water Saving (kL/Tonnic)
Recycle	Paper	0,84-1,00	6,33-10,76	0,60-3,20	2,91
Recycle	Glass	0,90-1,00	6,07-6,85	0,18-0,62	2,3
Recycle	Metal	0,79-0,96	36,09-191,42	1,40-17.8	5,97-81,77
Recycle	Plastic	0,90-0,97	38,81-64,08	0,95-1,88	-11,37
Recycle	Mixed Material	0,25-0,45	5,00-15,0	1,15	2,0-10
Composte	Organic Waste	0,60-0,65	0,18-0,47	0,25-0,75	0,44
Landfill	Mixed Waste	0	0,00.-0,84	(-)0,42-1,2	0

WASTE ELECTRIC AND ELECTRONIC EQUIPMENT (WEEE) IN ZERO WASTE STRATEGY

Waste Electric and Electronic Equipment (WEEE) (or e-waste) consists in different kinds of electronic and domestic devices such as refrigerators, air conditioners, cell phones, personal stereos, and consumer electronics to computers. The constituents of such goods contain toxic compounds such as mercury, chromium, arsenic, lead, cadmium, and plastics that require a special end-of-life treatment. The developed countries are the higher consumers of electronic goods, but the great problem is that such countries exported e-waste to the developing nations of Asia and Africa (as reported by StEP StEP. Initiative-Solving the E-waste problem. www.step-initiative.org). The e-waste production is almost increasing at the rate of 5 to 10% annually [62] and in the recent years such products are raising also in developing countries such as India and China. In the end-of-life stage of WEEE, the step of mechanical separation of re-useable components plays the key role in reducing the quantity of e-waste generated. According to a report from EPA, it is possible to save energy by recycling 1 million laptops and this value is equivalent to the electricity used by 3,657 U.S. houses in one year [63]. In this scenario, it is important to plan a green framework for WEEE waste management. In European Union nations, the legislations allow the free collection arrangements when consumers free return back their used e-waste, whereas in another developed nation such Japan, consumers pay an extra-charge when they return their electronic goods to the traders [64]. From March 2018, Singapore has deleted all batteries (including button cell batteries) containing more than 5 ppm by weight of mercury. Moreover, the manufacture, import and export of some products such as fluorescent lamps (exceeding specified mercury limits) and non-electronic measuring devices will be phased out by January 2020. In 2015, the National Environment Agency (NEA) formed a national voluntary partnership for e-waste recycling to promote proper recycling and treatment of discarded WEEE. Hereof, suitable bins have been located in government buildings, community clubs, schools, condominiums, shopping malls and major electronics retail stores. In each of them, a deposit slot was present to contain cables, mobile phones, tablets, laptops, DVD players, car stereos, telephones and answering machines. From 2012 to 2018, such project has collected more than 320 tonnes of e-waste [65]. The proper WEEE life cycle is quite complex (Fig. **4**) and in particular, the recycling chain consists of several subsequent steps: the first is the collection of e-waste. After the collection, such waste is sorted, dismantled and pre-processed in adequate plants and obtained materials must be subjected to proper treatment. First of all, we try to repair the products whenever possible, otherwise they are dismantled into their components and then separated into different categories. Such step consists of a mechanical separation into various fractions such as metals (iron, copper, aluminum *etc.*), plastics, ceramics, paper, wood and devices such as capacitors, batteries, picture tubes, LCDs, printed circuit boards *etc.* [66]. These separated segments can be used for new products or disassembled and sorted in different components such as toxic and non-toxic material. The toxic or hazardous components can be treated explicitly by the specific treatment, while non-toxic components can be processed. In this case, these components undergo different mechanical treatments (physical impaction, fragmentation/shredding and granulation) [67]. After going through the processes of size reduction, the leftover can be treated through the modes of solid waste handling. The 'green' framework for WEEE handling can increase both recycling rate and proper dismissal of e-waste. The Extended Producer Responsibility Framework Laws to Achieve Zero Waste' can support the 'green' framework for WEEE handling. Such laws state *that zero waste can be achieved through complete diversion of municipal solid waste (MSW) from* *landfills and incinerators, resource conversion, and sustainable product redesign.* European Union regulation (Directive 2002/95/EC, the RoHS- Restrictions on Hazardous Substances) provides the reduction of the use of hazardous substances in WEEE and also it proposes the reutilization and recycle of electronic equipments [68 - 70] This approach follows different aspects:

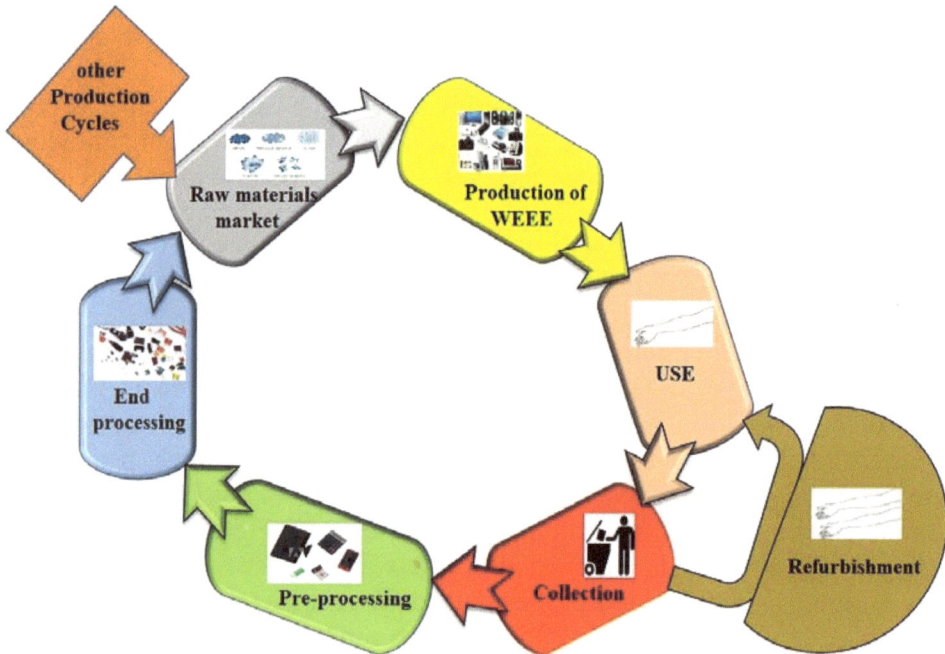

Fig. (4). The proper WEEE life cycle.

1. Lean Design: it consists of lean manufacturing to decrease the cost associated with applying an Environmental Management System.
2. Quality Control techniques (like Six Sigma strategy) can lead to fewer defects, diminished waste, fewer inputs required, and less energy.
3. Restrictions on Hazardous Substances: Limiting and restricting the use of hazardous and toxic substances in WEEE manufacturing.
4. Multi-purpose Design of electric and electronic equipment to decrease e-waste generation.
5. Green Manual is a new term that can be used in manuals of electronic instruments to explain to user about handling of devices after its end-of-life.

The proper WEEE life cycle in accordance with a 'zero waste' strategy is shown in Fig. (4).

Moreover, the more efficient management of WEEE is emphasized by the EU in 'A zero waste program for Europe' (COM 398, 2014). *'Zero waste' is one of the most visionary concepts for solving waste problems and seems to be a promising avenue for the management of waste in the future* [59].

ZERO WASTE PRACTICES ACROSS THE WORLD

The capital city of Ljubljana, Slovenia, became the first 'Zero Waste Capital' in Europe through new practices such as door-to-door separate collections, high composting and recycling rates [71]. Beside another town in Slovenia Vrhnika achieved the highest rate of recycling by 80% [72]. Italy is also one of the countries committed to zero waste and this has been proven since 232 cities are working towards this goal. Capannori, an Italian city, has reached up to 40% of waste reduction and 82% of waste segregation [72]. Capannori is one of the highest in Europe supported by good waste management policies and high public awareness [72]. In 1996, Canberra, Australia became the first regional government to adopt a Zero Waste strategy [73]. The strategy included resource recovery, recycling and reduction in consumption. In 2002 the New Zealand Government adopted The New Zealand Waste Strategy: Towards ZW and a Sustainable New Zealand. However, some papers detected that the zero waste endeavors had not been successful both in Canberra and in New Zealand. Furthermore, the more recent ACT Waste Management Strategy Towards a sustainable Canberra 2011–2025 [74] established a greater waste reduction to landfill, rather than total production system transformation. Similarly, in other study, Phillips *et al.* [75] noted that zero waste strategy in England was not associated with a real reduction of waste but only linked to waste minimization to landfill and incineration, Elsewhere, zero waste system have been implemented to increase recycling rates and to diminish total waste generation such as in Victoria (Australia) [76] and Taiwan [77]. Zotos *et al.* [78] observed that the adoption of zero waste policies regionally in Greece did not follow the policy targets of waste reduction as established by Canberra Zero Waste strategy. Other studies speculated that several organizations were mainly interested with reducing waste [79, 80] rather than to consider it as resource. For instance, in two articles, Davies [81, 82] suggested that the application of the policy at local level was weakly controlled by central government and, additionally, she also stressed the need of greater partnership of the private sector. At the end, she noted that the zero waste promotion was primarily been left to community groups without support of municipal, regional and governmental institutions.

In the United States, most of the places adopting zero waste goals are at the municipal county or regional level. According to the Zero Waste International

Alliance [83], California is the only state that has officially adopted zero waste goal. In a study of zero waste in Los Angeles, Murphy and Pincent [84] observed that waste management was disconnected with one of ecological aims for zero waste, namely, to consider the waste as a resource. Furthermore, they found that the city administration was more interested to increasing landfill diversion rates than the ecological consequences of this choice. In 2003, San Francisco set the objective of getting to zero waste by 2020. The first result of program "C40 Good Practice Guides: San Francisco - Zero Waste by 2020" was that the landfill disposal had a rate of 80% waste diversion in mid-2013 up from 35% in 1990. Today, through the number of measures and initiatives introduced, San Francisco has already reached a 78% recycling rate and has decreased the use of plastic bags by 100 million per year. Moreover, landfills reduction waste results in Greenhouse Gas (GHG) emissions savings, contributing of San Francisco's carbon decrease goal of 80% by 2050 (https://www.c40.org/case_studies/c40-good-practice-guides-san-francisco-zero-waste-by-2020) [85].

Different Cities (such as Oakland (CA), San Francisco (CA), Berkeley (CA) , Burbank (CA), Palo Alto (CA), Austin (TX), San Antonio (TX), Carrboro (NC), Seattle (WA), Kaual (HI)) and regions (such as (San louis Obisbo (CA),San Cruz County (CA), San Francisco County (CA), San Bernardino County (CA), San Diego County (CA), Sonoma County (CA), Boulder County (CO), Summit County (CO) and central Vermont Waste Management District) have adopted Zero Waste Strategy (as reported by Zero Waste International Alliance, Feb 16, 2013). As for the rest of the world, according to ZWIA, 25 international countries, regions, and cities have indicated adherence to zero waste goals. Australia and South Africa are two nations that have made zero waste commitments. In addition, the Regional Districts Nelson, Kootenay Boundary, Central Kootenay, Cowichan Valley, Sunshine Coast, and British Columbia have all made such commitments (from ZWIA, Feb 16, 2013).

ZERO WASTE PRACTICES IN DEVELOPING COUNTRIES

Most developing countries are not able to control solid waste management and their collection system is very irregular. Furthermore, such system varies from country to country and from town to town, and often the corruption at different levels (local and national) allows to create illegal waste management. Frequently, the municipalities of these countries cannot create a proper waste management system because of a lack of funds.

Today, zero waste has become an aspirational goal for developing countries to deal with their waste problems [50]. For instance, in Pune, India, waste pickers formed a union called '*KagadKach PatraKas htakari Panchayat*' [86]. These waste pickers demonstrated that communities could play a vital role to reach zero waste in order to live in sustainable way. Another study by Tangri [87] showed that a door-to-door collection service was present in Pune and operated by 2,000 waste pickers. In another paper, the author reported that the local government of Puerto Princesa (in the Philippines) has started zero waste program in association with a Non-governmental organization (NGO), the Mother Earth Foundation [88]. The aim of the project was to make Puerto Princesa a sustainable city by establishing a material recovery facility in every village and a waste separation center. The project was successful, since 90% of the residents did not mix waste and every village has at least one material recovery facility [88]. However, different studies have shown that many problems are involved in proper waste management in developing nations. The first thing is individuals' awareness that is important factor in determining the success of waste management [89]. In East Timor, a study observed that individuals did not sort their waste, because they did not have enough education and awareness in terms of treating the waste properly [90]. Another problem is the expansion of the population in the cities that results in an increase of waste generation due to the level of daily consumption [91], whereas landfill space is limited. For instance, one study, conducted in Sri Lanka by the Asian Institute of Technology, reported that the population were increased by 32.5% annually in association with an elevated waste generation. In this regard, in this nation, it is impossible to implement a sustainable waste management system (open dumping and burning waste) [92]. In Kenya, a study by Ziraba *et al.* [93] pointed out the association between poor waste management and adverse health outcomes. A research study by Al-Khatib *et al.* [94] observed that open dumps was still utilized to dispose waste in seven Palestinian districts. In recent years, many developing countries struggle with the amount of plastic waste [95]. A recent report stated that China, Indonesia, the Philippines, Vietnam, and Thailand ranked 1 to 5 respectively in plastic waste generation into the oceans [96]. Around 40% of plastic waste originated from packaging products as a result of unsustainable consumption patterns [97]. However, there are few studies on zero waste around the world: Australia (12), USA (12), South Africa (7), England (5), Taiwan (5) India (5), Sweden (4), China (3), Japan (3), Thailand (3) and Brazil (3) [27]. Future research should be conducted as a part of a long-term monitoring and evaluation system. For example, it is crucial to understand the relationship between zero waste experience and communities' ability to attain a real sustainable waste management. Furthermore, these examples should be periodically examined at different locations to monitor the obtained changes. Additional follow-up research should be conducted to identify which changes in waste management are considered important by the communities.

CONCLUSIONS

The waste management system represents an important component of actions for achieving UN-SDGs. It is legally under the control of the governments that give little priority to this problem [98]. As a consequence, the lack of monitoring and the evaluation of waste management, the low enforcement of regulations, and limited financial resources became major drawbacks for the government to support a sustainable waste management [99, 100]. Many authors reported that insufficient infrastructure, poor roads and limited vehicles for waste collection resulted in an improper waste collection system [101, 102]. Then, the improvement of waste collection services, the elimination of open dumping and burning, and the upgrading dumpsites are the initial steps in terms of the SDGs in developing countries. Today, more than 50% of the world population is living in big cities where there is a high consumption rate. For this reason, the throwaway society concept has to change, and zero waste concept has to be developed and improved among people to protect our planet from deterioration and to handover it clean and green to our future generations. The consumption-based model is becoming unsustainable above all in the 2050, when the human population will potentially reach 9 billion persons [103]. Once again, the role of government will be significant through policy support and incentives and sanctions for waste management above all for the industry sector. Today, zero waste is the best possible alternative in solving the problems of waste generation. A change of our current economic growth model will be necessary to go towards a green economy that opens us to many opportunities as well as posing many challenges' [104, 105]. Furthermore, new interplay among local government, national government, industry and community is necessary to better handle the step from waste management to materials management.

The challenges are:

- Traditional governance for waste management is at municipality level. In this case, waste management is considered only as a cost rather than a growth opportunity for new companies.
- Central governments must drive the different system of waste management with policies that aim to develop new infrastructures for materials recovery.
- Industry involvement in materials recovery could play a crucial role in the alternative methods of waste management through new commercialization models.
- Grassroots community and business initiatives such as, transition and eco-towns, design innovations, need the support of central governments for production, consumption and materials management sustainability.

Economic incentives should be provided by governments for those industries that use materials already present in the production system to encourage closed loop production benefits.

In this context, it is necessary:

- Association between national resource management strategies with waste and materials and new policy making.
- Increase of producer responsibility to better recover material assets lost to the supply-chain and to reduce long-term costs.
- From a policy perspective, to incentive sustainable consumption behavior through policies to discourage excessive consumption.

In conclusion, zero waste strategy is based on socioeconomic changes and innovative approaches not only to resources and production but above all to consumption, which is very hard in today's society. Such process could be fostered by a good policy coordination of all the actions needed for the application of SDG's. This involves both a horizontal level, among policies (in particular, macroeconomic, industrial, research and innovation, environmental) and a vertical level, among institutions (supranational, national, regional, local and municipal). (https://m.facebook.com/story.php?story_fbid=1240948522756120&id =464101943774119).

CONSENT FOR PUBLICATION

Not applicable.

CONFLICT OF INTEREST

The author declares that there is no conflict of interest in this chapter.

ACKNOWLEDGEMENTS

Declared none.

REFERENCES

[1]   WCED. U Our common future. World Commission on Environment and Development Oxford University Press 1987.

[2]   Elliott J. An Introduction to Sustainable Development. 4th ed., London: Routledge 2012. [http://dx.doi.org/10.4324/9780203844175]

[3]   2002.http://www.undocuments.net/aconf199-20.pdf

[4]   El-Haggar S. Sustainable Industrial Design and Waste Management: Cradle To-cradle for Sustainable Development. Sheffield: Academic Press 2010.

[5]   Tudor T, Robinson GM, Riley M, Guilbert S, Barr SW. Challenges facing the sustainable consumption and waste management agendas: perspectives on UK households. Local Environ 2011; 16(1): 51-66.

[http://dx.doi.org/10.1080/13549839.2010.548372]

[6]   Farrelly T, Schneider P, Stupples P. Trading in waste: Integrating sustainable development goals and environmental policies in trade negotiations toward enhanced solid waste management in Pacific Islands countries and territories. Asia Pac Viewp 2016; 57(1): 27-43.
[http://dx.doi.org/10.1111/apv.12110]

[7]   Gupta S, Dangayach GS. Sustainable waste management: a case from Indian cement industry. Braz J Oper Prod Manag 2015; 12(2): 270-9.
[http://dx.doi.org/10.14488/BJOPM.2015.v12.n2.a7]

[8]   United Nations. Transforming Our World: The 2030 Agenda for Sustainable Development. New York: UN. 2015.

[9]   Cooper D, Timothy TG. The environmental impacts of reuse: A review. J Ind Ecol 2015.
[http://dx.doi.org/10.1111/jiec.12388]

[10]   Rodic L, Wilson DC. Resolving governance issues to achieve priority sustainable development goals related to solid waste management in developing countries. Sustainability 2017; 9: 404.
[http://dx.doi.org/10.3390/su9030404]

[11]   2016.https://doi.org/978-2-940529-40-7

[12]   Van Cauwenberghe L, Janssen CR. Microplastics in bivalves cultured for human consumption. Environ Pollut 2014; 193: 65-70.
[http://dx.doi.org/10.1016/j.envpol.2014.06.010] [PMID: 25005888]

[13]   Pauly JL, Stegmeier SJ, Allaart HA, *et al.* Inhaled cellulosic and plastic fibers found in human lung tissue. Cancer Epidemiol Biomarkers Prev 1998; 7(5): 419-28.
[PMID: 9610792]

[14]   Rhodes CJ. Plastic pollution and potential solutions. Sci Prog 2018; 1;101(3): 207-60.
[http://dx.doi.org/10.3184/003685018X15294876706211]

[15]   United Nations Human Settlements Programme (UN-HABITAT). Collection of municipal waste in developing countries. Gutenberg Press: Malta 2010.

[16]   Ngoc UN, Schnitzer H. Sustainable solutions for solid waste management in Southeast Asian countries. Waste Manag 2009; 29(6): 1982-95.
[http://dx.doi.org/10.1016/j.wasman.2008.08.031] [PMID: 19285384]

[17]   Wilson DC. Development drivers for waste management. Waste Manag Res 2007; 25(3): 198-207.
[http://dx.doi.org/10.1177/0734242X07079149] [PMID: 17612318]

[18]   Sakai S, Yoshida H, Hirai Y, *et al.* International comparative study of 3R and waste management policy developments. J Mater Cycles Waste Manag 2011; 3: 86-102.
[http://dx.doi.org/10.1007/s10163-011-0009-x]

[19]   Gertsakis J, Lewis H. Sustainability and the Waste Management Hierarchy: A Discussion Paper on the Waste Management Hierarchy and its Relationship to Sustainability. Melbourne: RMIT University 2003; pp. 1-15.

[20]   Ferrara I, Missios P. Does Waste Management Policy Crowd Out Social and Moral Motives for Recycling. Working Paper 031, Department of Economics, Ryerson University: Mimeo 2012.

[21]   Davies A, Fahy F, Taylor D, Meade H. Environmental attitudes and behaviour: values, actions and waste management. Environmental RTDI Programme 2000-2006, Final report 2005.

[22]   Townsend TG, Powell J, Jain P, Xu Q, Tolaymat T, Reinhart D. Planning for sustainable landfilling practices Chapter: sustainable practices for landfill design and operation. Waste Management. Princ Pract 2015; pp. 35-51.
[http://dx.doi.org/10.1007/978-1-4939-2662-6_3]

[23]   Ferronato N, Torretta V. Waste mismanagement in developing countries: A review of global issues.

Int J Environ Res Public Health 2019; 16(6): 1982-95.
[http://dx.doi.org/10.3390/ijerph16061060] [PMID: 30909625]

[24] Al-Khatib IA, Sato C. Solid health care waste management status at health care centers in the West Bank--Palestinian Territory. Waste Manag 2009; 29(8): 2398-403.
[http://dx.doi.org/10.1016/j.wasman.2009.03.014] [PMID: 19398317]

[25] Coker A, Sangodoyin A, Sridhar M, Booth C, Olomolaiye P, Hammond F. Medical waste management in Ibadan, Nigeria: obstacles and prospects. Waste Manag 2009; 29(2): 804-11.
[http://dx.doi.org/10.1016/j.wasman.2008.06.040] [PMID: 18835151]

[26] Patwary MA, O'Hare WT, Sarker MH. An illicit economy: scavenging and recycling of medical waste. J Environ Manage 2011; 92(11): 2900-6.
[http://dx.doi.org/10.1016/j.jenvman.2011.06.051] [PMID: 21820235]

[27] Zaman AU. A comprehensive review of the development of zero waste management: lessons learned and guidelines. J Clean Prod 2015; 91: 12-25.
[http://dx.doi.org/10.1016/j.jclepro.2014.12.013]

[28] 2004.Zero waste definition Retrieved December 20, 2012; from Zero Waste International Alliance: zwia.org/standards/zw-definition

[29] Palmer P. Getting to Zero Waste: Universal Recycling as a Practical Alternative to Endless Attempts to" Clean up Pollution. Portland: Purple Sky Press 2011.

[30] Zaman AU, Lehmann S. Challenges and opportunities in transforming a city into a "zero waste city". Challenges 2011; 2(4): 73-93.
[http://dx.doi.org/10.3390/challe2040073]

[31] Mary Lou Derventer personal communication. 2012.

[32] Cradle to Cradle: Remaking the Way We Make Things. New York: North point press 2002.

[33] http://www.cleanproduction.org/Publications.php

[34] http://www.productpolicy.org/resources/index.html

[35] http://www.grrn.org/zerowaste/business/index.phpand
http://www.grrn.org/zerowaste/business/profiles.php
http://66.35.240.8/cgibin/article.cgi?f=/c/a/2007/07/20/BAGCSQV2661.DTL

[36] San Francisco Zero waste program see videotape: "On the Road to Zero Waste, part 4: San Francicso, produced by Paul Connett for GG video, 2004, 29 minutes, GG Video, 82 Judson Street, Canton, NY 13617. and see SFenvironment web page.

[37] Appelhof M. Worms Eat My Garbage. Kalamazoo, Michigan: Flower Press 1982.

[38] Rosenthal E. Motivated by a Tax, Irish Spurn Plastic Bags 2008. http://www.nytimes.com/2008/02/02/world/europe/02bags.html

[39] Lowy J. Plastic left holding the bag as environmental plague. Nations around world look at a ban. Seattle Post-Intelligencer 2004; 21(July)http://www.commondreams.org/headlines04/0721-04.htm

[40] Goodyear C. 2007. http://www.sfgate.com/cgibin/article.cgi?file=/c/a/2007/03/28/MNGDROT5QN1.DTL http://www.grillibiellesi.org http://www.federicopistono.org

[41] Gourmelon G. 2015.vitalsigns.worldwatch.org

[42] 2019.http://worldpopulationreview.com/continents/africa-population/

[43] 2019.https://www.newplasticseconomy.org/assets/doc/Global-Commitment-2019-Progress-Report.pdf

[44] The Beer Store featured in the video "Target Zero Canada" produced by Paul Connett for GG video, Jan 01, 52 minutes, GG Video, 82 Judson Street, Canton, NY 13617 2019.

[45] Connett, P. Zero Waste: A Key Move Towards A Sustainable Society, American Environmental

Health Studies Project (USA: Canton) 2007. (https://www.researchgate.net/publication/228871831_Zero_Waste_A_Key_Move_towards_a_Sustainable_Society)

[46]   North R. http://www.recyclenorth.org/

[47]   Recycle North featured in the video. On the Road to Zero Waste, part 2: Burlington, Vermont produced by Paul Connett for GG video 2001 Jan; length, 29 minutes, GG Video, 82 Judson Street, Canton, NY 13617.

[48]   Urban Ore featured in the video. Zero Waste: Idealistic Dream or Realistic Goal? produced by Paul Connett for GG video, Sept 1999. length 58 minutes, GG Video, 82 Judson Street, Canton, NY 13617.

[49]   Sonrise Recycling featured in the video. Zero Waste: Idealistic Dream or Realistic Goal? produced by Paul Connett for GG video, Sept 1999. length 58 minutes, GG Video, 82 Judson Street, Canton, NY 13617.

[50]   Renovators Resource featured in the video. On the Road to Zero Waste, part1: Nova Scotia. produced by Paul Connett for GG video 2001. length 29 minutes, GG Video, 82 Judson Street, Canton, NY 13617.

[51]   Knox R. Towns tilting to pay-per-bag trash disposal 2007. http://www.boston.com/news/local/articles/2007/11/08/towns_tilting_to_pay_per_bag_trash_disposal/

[52]   Hanley R. Pay-by-Bag Trash Disposal Really Pays, Town Learns 1988. http://query.nytimes.com/gst/fullpage.html?res=940DE7DA123CF937A15752C1

[53]   Videotape: "On the Road to Zero Waste, part 1: Nova Scotia" produced by Paul Connett for GG video, 2001, 29 minutes, GG Video, 82 Judson Street, Canton, NY 13617 2001.

[54]   Nova Scotia program, for a governmental perspective contact Barry Friessen, barry. 2019.

[55]   Nova Scotia program, for a citizen's perspective contact David Wimberly 2019.

[56]   Nizar M, Munir E, Munawar E, Matseh I, Waller V. Applying Zero Waste Management Concept in a City of Indonesia: A Literature Review International. Int J Eng Technol, 2018, 7 (4), 6072-6077.

[57]   Yoshida H, Gable JJ, Park JK. Evaluation of organic waste diversion alternatives for greenhouse gas reduction. Resour Conserv Recycling 2012; 60: 1-9.
[http://dx.doi.org/10.1016/j.resconrec.2011.11.011]

[58]   Mueller W. The effectiveness of recycling policy options: waste diversion or just diversions? Waste Manag 2013; 33(3): 508-18.
[http://dx.doi.org/10.1016/j.wasman.2012.12.007] [PMID: 23312779]

[59]   Zaman AU, Lehmann S. The zero waste index: a performance measurement tool for waste management systems in a 'zero waste city'. J Clean Product 2013; 50: 123-32.

[60]   Mazzanti M, Montini A, Nicolli F. The dynamics of Landfill Diversion: Economic Drivers, Policy Factors and Spatial Issues: Evidence from Italy Using Provincial Panel Data. Resour Conserv Recycling 2009; 54: 53-61.
[http://dx.doi.org/10.1016/j.resconrec.2009.06.007]

[61]   Kasam FMI, Satrio AP. Evaluation of solid waste management at campus using the "Zero Waste Index": The case on campus of Islamic University of Indonesia MATEC Web of Conferences 154, 02004 ICET4SD 2017 2018.
[http://dx.doi.org/10.1051/matecconf/201815402004]

[62]   Sthiannopkao S, Wong MH. Handling e-waste in developed and developing countries: initiatives, practices, and consequences. Sci Total Environ 2013; 463-464: 1147-53.
[http://dx.doi.org/10.1016/j.scitotenv.2012.06.088] [PMID: 22858354]

[63]   Frequent Questions | eCycling. EPA , [Accessed March 2, 2014.];

[64]   Khan SS, Lodhi SA, Akhtar F, Khokar I. Challenges of waste of electric and electronic equipment (WEEE). Challenges of waste of electric and electronic equipment (WEEE). Manag Environ Qual

2014; 25: 166-85.

[65]   https://www.reach.gov.sg/participate/public-consultation/national-environment-agency/pollut-on-control/tightening-of-controls-for-mercury-added-batteries-and-button-cellbatteries

[66]   Gramatyka P, Nowosielski R, Sakiewicz P. Recycling of waste electrical and electronic equipment. J Achiev Mater Manuf Eng 2007; 20: 535-8.

[67]   Dalrymple I, Wright N, Kellner R, *et al.* An integrated approach to electronic waste (WEEE) recycling. Circuit World 2007; 33(2): 52-8.
[http://dx.doi.org/10.1108/03056120710750256]

[68]   Austin AA. Where will all the waste go? utilizing Extended Producer Responsibility Framework Laws to achieve zero waste. Golden Gate University Law Journal 2014 March 3;

[69]   Austin AA. 8 Directive 2002/96/EC of the European Parliament and of the Council of the 27 January 2003 on waste of electrical and electronic equipment. J EU L3 2003; 24-38.

[70]   Jaiswal A, Patel CSBS. Manish Kumar M Go Green with WEEE: Eco-friendly approach for handling e- waste. Procedia Comput Sci 2015; 46: 1317-24.
[http://dx.doi.org/10.1016/j.procs.2015.01.059]

[71]   Ilić M, Nikolić M. Drivers for development of circular economy–A case study of Serbia. Habitat Int 2016; 56: 191-200.
[http://dx.doi.org/10.1016/j.habitatint.2016.06.003]

[72]   Van Vliet A. Zero Waste Europe Case Study 1: The Story of Capannori. In: Europe ZW, Ed. The Story of Capannori. Netherlands: Zero Waste Europe 2014.

[73]   Australian Capital Territory (ACT) Government 1996.

[74]   ACT Waste Management Strategy Reducing waste and recovering resources to achieve a sustainable, carbon-neutral Canberra 2011–2020.

[75]   Phillips PS, Tudor T, Bird H, Bates MA. Critical Review of key waste strategy initiative in England: Zero Waste Places Projects 2008-2009 Resources. Conservation and Recycling 2011; 55(3): 335-43.
[http://dx.doi.org/10.1016/j.resconrec.2010.10.006]

[76]   Clay S, Gibson D, Ward J. Sustainable victoria: Influencing resource use, towards zero waste and sustainable production and consumption. J Clean Product 2007; 15(8-9): 782-6.

[77]   Young CY, Ni SP. Working towards a zero waste environment in Taiwan. Waste Manag Res 2010; 28: 236-2424.

[78]   Zotos G, Karagiannidis A, Zampetoglou S, *et al.* Developing a holistic strategy for integrated waste management within municipal planning: challenges, policies, solutions and perspectives for Hellenic municipalities in the zero-waste, low-cost direction. Waste Manag 2009; 29(5): 1686-92.
[http://dx.doi.org/10.1016/j.wasman.2008.11.016] [PMID: 19147341]

[79]   Watson M, Bulkeley H. Just waste? Municipal waste management and the politics of environmental. Just Local Environ 2005; 10(4): 411-26.

[80]   Watson M, Bulkeley H, Hudson R. Unpicking Environmental Policy Integration with Tales rrom Waste Management Environment and Planning C: Government and Policy 2008; 26(3): 481-98.

[81]   Davies AR. Geographies of Garbage Governance: Interventions, Interactions & Outcomes Aldergate. Ashgate Publishing Limited 2008.

[82]   Davies AR. Clean and green? A governance analysis of waste management in New Zealand. J Environ Plann Manage 2009; 52: 157-78.
[http://dx.doi.org/10.1080/09640560802666503]

[83]   Liss G. Zero waste communities 2013.http://zwia.org/news/zerowaste-communities

[84]   Murphy S, Pincetin S. Zero Waste in Los Angeles: Is the Emperor Wearing Any Clothes Resources.

Conserv Recycl 2013; 81: 40-51.
[http://dx.doi.org/10.1016/j.resconrec.2013.09.012]

[85]   2016.www.calstatela.edu/sites/default/files/groups/Sustainability/cal_state_la_zero_waste_plan.pdf

[86]   Chikarmane P, Narayan L. Formalising Livelihood: Case of Waste pickers in Pune. Econ Polit Wkly 2000; 35(41): 3639-42.

[87]   Tangri N. Pune. India: Waste Pickers Lead the Way to Zero Waste 2012.https://www.no-burn.org/w--content/uploads/ZW-Pune.pdf88

[88]   Atienza V. Review of the Waste Management System in the Philippines: Initiatives to Promote Waste Segregation and Recycling Through Good Governance.Economic Integration and Recycling in Asia: An Interim Report. Tokyo: Institute of Developing Economies 2011; pp. 65-97.

[89]   Babaei AA, Alavi N, Goudarzi G, Teymouri P, Ahmadi K, Rafiee M. Household recycling knowledge, attitudes and practices towards solid waste management. Resour Conserv Recycling 2015; 102: 94-100.
[http://dx.doi.org/10.1016/j.resconrec.2015.06.014]

[90]   Soares J, Yunus HS, Kusuma D. Public perception of urban solid waste management in sub district domaleixo Dili-Timor Leste. Geogr Indonesia Mag 2016; 25(2): 162-80.

[91]   Farrelly T, Tucker C. Action research and residential waste minimisation in Palmerston North, New Zealand. Resour Conserv Recycling 2014; 91: 11-26.
[http://dx.doi.org/10.1016/j.resconrec.2014.07.003]

[92]   Bandara NJGJ. Municipal solid waste management The Srilankan case in developmens in Forestry and Environment Management in Sri Lanka. Department of Forestry and Environmental Sciences: University of Sri Jayewardenepura 2008; pp. 93-5.

[93]   Ziraba AK, Haregu TN, Mberu B. A review and framework for understanding the potential impact of poor solid waste management on health in developing countries. Arch Public Health 2016; 74(1): 55.
[http://dx.doi.org/10.1186/s13690-016-0166-4] [PMID: 28031815]

[94]   Al-Khatib IA, Arafat HA, Basheer T, *et al.* Trends and problems of solid waste management in developing countries: a case study in seven Palestinian districts. Waste Manag 2007; 27(12): 1910-9.
[http://dx.doi.org/10.1016/j.wasman.2006.11.006] [PMID: 17224264]

[95]   Shekdar AV. Sustainable solid waste management: an integrated approach for Asian countries. Waste Manag 2009; 29(4): 1438-48.
[http://dx.doi.org/10.1016/j.wasman.2008.08.025]

[96]   Jambeck JR, Geyer R, Wilcox C, *et al.* Marine pollution. Plastic waste inputs from land into the ocean. Science 2015; 347(6223): 768-71.
[http://dx.doi.org/10.1126/science.1260352] [PMID: 25678662]

[97]   Comaniţa ED, Hlihor RM, Ghinea C, Gavrilescu M. Occurrence of plastic waste in the environment: ecological and health risks. Environ Eng Manag J 2016; 15(3): 675-85. [EEMJ].
[http://dx.doi.org/10.30638/eemj.2016.073]

[98]   Guerrero LA, Maas G, Hogland W. Solid waste management challenges for cities in developing countries. Waste Manag 2013; 33(1): 220-32.
[http://dx.doi.org/10.1016/j.wasman.2012.09.008]

[99]   Hazra T, Goel S. Solid waste management in Kolkata, India: practices and challenges. Waste Manag 2009; 29(1): 470-8.
[http://dx.doi.org/10.1016/j.wasman.2008.01.023] [PMID: 18434129]

[100] Alavi Moghadam MR, Mokhtarani N, Mokhtarani B. Municipal solid waste management in Rasht City, Iran. Waste Manag 2009; 29(1): 485-9.
[http://dx.doi.org/10.1016/j.wasman.2008.02.029] [PMID: 18448322]

[101] Henry RK, Yongsheng Z, Jun D. Municipal solid waste management challenges in developing

countries--Kenyan case study. Waste Manag 2006; 26(1): 92-100.
[http://dx.doi.org/10.1016/j.wasman.2005.03.007] [PMID: 16006111]

[102] Aleluia J, Ferrão P. Characterization of urban waste management practices in developing Asian countries: A new analytical framework based on waste characteristics and urban dimension. Waste Manag 2016; 58: 415-29.
[http://dx.doi.org/10.1016/j.wasman.2016.05.008] [PMID: 27220609]

[103] Emmott S. 10 Billion London Penguin 2013.

[104] Jackson T. Prosperity without growth? The transition to a sustainable economy. Sustainable Development Commission 2009.

[105] Jackson T, Senker P. Prosperity without growth: economics for a finite planet. Energy Environ 2011; 22: 1013-6.
[http://dx.doi.org/10.1260/0958-305X.22.7.1013]

SUBJECT INDEX

A

Abnormal 79, 84, 130
  chromatin structure 130
  heart rhythm 79
  Ras oncogene 84
Abnormalities 68, 78, 83, 85, 130, 147, 148
  behavioral 148
  chromosomal 68, 130
  human congenital 85
Absorption 56, 67, 71, 72, 126
  nutrient 71
  pathway 126
  selective 72
Acids, polyunsaturated fatty 131
Activities 19, 97, 99, 100, 122, 124
  anthropogenic 97, 99
  chemical 19
  enzymatic 124
  enzyme 100, 122
  metabolic 100
Advanced waste treatment (AWT) 178
Agencies 7, 43, 147, 154
  regulatory 147, 154
  specialized international 7
Agency for toxic substances and disease
      registry (ATSDR) 38, 120
Agents 63, 67, 70, 72, 76, 90, 98, 115, 150
  biological 150
  chemical 76
  flocculating 98
  non-carcinogenic 63
Agriculture 5, 53, 116, 147, 166, 163
  intensive 116
  sustainable 163, 166
Air 10, 27, 30, 31, 35, 76, 99, 133, 179
  concentration 10
  conditioners 179
  emissions 27, 30, 31
  inhalation 99
  measurement 35
  pollution 76, 133
Allergic reactions 65, 71, 79, 80

Alterations 101, 115, 122, 123, 124, 127, 129,
    130, 131, 133
  epigenetic 129
  genes 101
  genetic 124
  molecular 115
  pathological 122
Alzheimer's disease 78
Anaerobic digestion 97, 101
Analysis 37, 88
  fungi 88
  geospatial 37
Aneuploidy 130
  chromosomal 130
Anomalies 36, 37, 62, 97
  congenital 36, 37, 62
  developmental 97
Antioxidant enzymes 86
Apoptosis 132
Application, nuclear technology 76
Asthma 78, 80
Atherosclerosis 124
Auto-oxidation reactions 86

B

Biological Effects 62, 82, 116, 126
  detrimental 116
  toxic 62
Bird movement modelling 39
Bladder cancer 34, 36, 38, 41, 76, 115
  urinary 115
Blood circulation 66
Brumadinho dam disasters 151

C

Cadmium 75, 101, 103
  absorption 101
  consumption 101
  electrode 103
  fumes 75

Cancer 5, 6, 7, 8, 9, 34, 35, 36, 37, 38, 67, 68, 69, 70, 71, 84, 115, 127, 128, 150
  brain 34
  breast 35, 38, 115
  childhood 37
  developing 34
  excess 39
  glandular 36
  hematologic 38
  hepato-biliary 34
  laryngeal 36
  nasopharyngeal 115
  non-metastatic 67
  ovarian 115
  prostate 115
  testicular 127, 128
  thyroid 150
Cancer risks 3, 34, 35, 37, 38, 71, 84, 85
  bladder 38
  excess 37
Carcinogenesis 63, 72, 100, 115, 122, 123
Carcinogenicity 2, 22, 63, 70
Carcinogens 4, 22, 63, 69, 70, 72, 79, 80, 84, 115
  environmental 115
  non-gene 63
  non-toxic 69
  suspected 4, 79
Carcinoma 74, 115
  hepatocellular 115
  papillary thyroid 115
Cardiorespiratory diseases 36
Cardiovascular disease 67
Chemicals 38, 57, 65, 68, 69, 77, 169
  carcinogenic 57, 69
  synthetic 77
  toxic 38, 65, 68, 77, 169
Chronic 38, 79, 133
  beryllium disease 79
  diseases prevention 133
  lymphatic leukemia (CLL) 38
Cirrhosis 68, 71
Coal 151
  ash slurry 151
  combustion 151
  fly ash slurry 151
Collection systems 25, 42, 174, 176, 183, 185
  door-to-door 174
  improper waste 185
  started waste 42

Combustible waste petroleum 55
Comprehensive environmental response, compensation and liability act (CERCLA) 4, 54, 120
Contaminants 2, 3, 8, 37, 56, 59, 76, 115, 117, 120, 121, 122, 123, 124, 125, 126, 127, 131, 133
  chemical 127
  dangerous 2, 3
  dangerous waste 76
  environmental 121, 129, 131
  lipophilic 37
  physical 126
Cryptorchidism 128
Cyanide 73, 74
  complex 74
  free 74
  hydrogen 74
Cyanide ions 74

D

Damage 4, 68, 70, 78, 79, 80, 81, 114, 121, 122, 126, 127, 131, 134
  chromosome 4
  fetal 80
  genetic 4
  molecular levels 122
  spermiogenesis 131
Deaths 4, 7, 9, 64, 65, 68, 71, 73, 77, 150, 151, 152, 153, 154
  cell 73
  increased fetal 77
Defects 37, 130, 181
  genomic material 130
  neural-tube 37
  sperm DNA integrity 130
Deformities 68, 72, 97
  congenital 68
Degenerative diseases 127
  chronic 133, 134
Depression 27, 146, 150, 153, 155
  natural 27
Differences 82, 117, 118, 128, 132
  genetic 128
Disabilities 154, 168
  intellectual 154
Disability-adjusted life year (DALYs) 40, 41
Disaster 152, 153, 154
  nuclear 153, 154

Disease induction 84
Disease risk 124
Diseases 34, 99, 113, 115, 116, 124
  chronic-degenerative 113, 116
  environmental 124
  metabolic 115
  respiratory 34, 99
  transmitted 113
Disorders 67, 75, 83, 84, 125, 128
  biological 84
  complex prevalent 83
  immune 125
  sexual development 67
Dispersion modelling 35
Disposal 17, 18, 23, 24, 25, 43, 51, 57, 58, 64,
    85, 88, 99, 172
  dangerous 172
  of hazardous wastes 17, 43
DNA 84, 90, 114, 124, 127, 129, 130
  biomolecular level 114
  deoxyribonucleic 84
    helix grooves 129
    hypermethylation 129
DNA damage 125
  repair mechanisms 125
DNA fragmentation 130
  chromosomal 130
  defects 130
Drugs 77, 78, 129, 150
  cancer chemotherapy 77
  nausea-preventing 78
Dump 4, 5, 10
  factories 5
  industrial 4
Dumping 6, 10, 22, 24, 88, 167
  illegal 6, 22, 24

E

Ecological resources 119, 120, 121
  existing 119
Economic expansion 51, 149
  sustainable 149
Economic growth 4, 5, 149, 159, 160, 163,
    168
  rapid 4, 5
  sustainable 163, 168
Ecosystems 37, 119, 123, 159, 167, 169
  inland freshwater 169

  surrounding 37
  water-related 167
Effects 2, 4, 65, 66, 68, 76, 81, 84, 86, 87, 98,
    114, 115, 131, 133
  antagonistic 65
  cellular genetic 84
  deficiency 115
  destructive 87
  increased greenhouse gas 4
  nephrotoxic 81
  non-cancer 2
  non-reversible 68
  of hazardous wastes 86
  of heavy metals on plants 98
  reversible 68
  synergistic 76
  systemic 66
  therapeutic 65
  transgenerational 114, 131, 133
Efficiency 71, 98, 103, 168, 179
  global resource 168
  increased floatation 103
Effluent, municipal waste-water 8
Electrodialysis 101
Emissions 6, 24, 28, 34, 35, 51, 132, 177, 178,
    179
  atmospheric 51
  gaseous 24
  high dioxin 34, 35
  increased dioxin 35
Endothelial nitric oxide synthase 86
Energy 88, 99, 161, 166, 167, 168, 172, 177,
    180, 181
  resilient 168
  sustainable 161
Environmental 1, 121, 151, 176
  contamination 1, 121
  degradation-pollution, massive 151
  epidemiology 121
  issues 176
Environmentally sound management (ESM) 1,
  166
Environmental 4, 33, 37, 39, 71, 113, 116,
    118, 120, 121, 122, 123, 147, 148, 153,
    171, 180, 181
  management system 181
  monitoring programs 122
  movement 171
  pressures 113, 116, 118

protection agency (EPA) 4, 33, 37, 39, 120,
    121, 147, 148, 153, 180
  quality studies 123
  risk areas 113
  toxicology 71
EPA test method 21, 22
Epigenetic damages 129
Estrogen receptors 67
Excretion 71, 83, 101
  urinary 83, 101
Exposure 2, 3, 8, 21, 34, 35, 36, 41, 56, 62,
    64, 66, 67, 68, 69, 74, 79, 81, 82, 83, 84,
    85, 100, 123, 124, 126, 146, 147, 152,
    153, 154
  accidental 8, 62
  acute 56, 66
  aromatic hydrocarbon 81
  chemical 83, 153
  chronic 56, 66, 74, 79, 84
  dermal 21
  effects of hazardous waste and radiation
    153, 154
  nephrotoxic 81
  prolonged 62
  radiation 152, 153, 154
  systematic 152
  toxic waste 146, 147
Exposure pathways 2, 3, 63, 126
  human 63
  test 63

F

Factors 54, 59, 63, 65, 66, 68, 69, 70, 76, 82,
    85, 115, 124, 127, 128, 130, 149
  atmospheric 68
  behavioral 127
  biological 82, 149
  environmental 85, 115, 124, 127, 128, 149
  environmental exposure 124
  genetic 82, 128
  genetic susceptibility 115
  hydrogeological 63
  intrinsic 115
Faculty of civil engineering and planning
    (FCEP) 178
Female reproductive cycle 77
Fertility 6, 77, 113, 116, 127, 128, 130, 131,
    133, 154
  assessment 133

feminine 77
  male 127, 130, 131
  problems 113
  reduced 116
Filtration 97, 102, 103
  membrane 102
  methods 102
Fishermen's erythrocytes 132
Food 67, 68, 75, 97, 98, 99, 100, 101, 127,
    128, 161, 175, 178, 179
  contamination 75
  staple 101
Food waste 166
  reduction 166
Fractions 71, 172, 174, 176, 180
  clean organic 174
  residual 176
Free radical formation of heavy metals 100

G

Garbage 4, 178
  decomposing 4
Gases 2, 4, 19, 21, 25, 30, 71, 74, 88, 97, 98,
    99, 103, 147, 152, 169, 177, 178, 179,
    183
  containerized 19
  greenhouse 169, 177, 178, 179, 183
  industrial vent 99
  toxic 2, 74, 152
Gene promoters 129
Genes 63, 83, 84, 101, 115, 124, 125, 128,
    129, 130
  key 115
  overexpressed 84
  toxic 63
  tumor 84
Genetic 63, 85, 124
  developmental 63
  diseases 63
  polymorphism 124
  predisposition 85
Genomic instability 72, 90, 124
Genotoxic 3, 4, 64, 72, 90, 131
  agents 4
  interaction 3
Genotoxicity 56
Geographic information systems (GIS) 31, 33,
    35
Global-warming pollution 10

Glucuronic acid 72
Glutathione 4, 86, 114, 115, 169, 177, 178, 179, 183
  gas (GHG) 4, 169, 177, 178, 179, 183
  reduced 114
  peroxidase 86
  reductase 114
Groundwater 4, 25
  contaminations 25
  pollution 4
GSH-reductase 86

H

Hazardous 17, 20, 31, 41, 56, 63, 64, 78, 80, 88, 147, 150, 166, 167, 180
  chemicals 20, 41, 64, 166, 167
  components 56, 180
  compounds 56
  crises and radioactive accidents 150
  facilities 147
  gaseous substances 31
  household wastes 17
  materials 63, 64, 88
  Substances in PCs 78, 80
Hazardous waste 4, 7, 19, 23, 24, 33, 52, 53, 55, 56, 57, 58, 62, 81, 146, 148
  amount of 53, 55, 57, 58
  collocate 24
  complex nature of 52
  defined 19
  dispose 57
  exposure to 56, 62, 81, 148
  generation of 4, 7, 23, 33, 55, 56
  handling 52
  industrial 33, 53
  managing 52
  negative impact of 146
Hazardous waste facility site 32
  selection 32, 33
Hazardous Waste Landfill(s) 25, 29, 30
  disposal 25, 29, 30
Hazardous waste management 39, 41, 57, 62
  business 41
  decision 62
  systems 39, 57
Health effects 9, 34, 78, 79, 80, 81, 84, 97, 100, 106, 113, 121, 150, 153
  adverse 9, 78, 79, 80, 81, 113
  embryonic toxicological 100

  mental 150, 153
  pathways 121
Health impacts 3, 8, 56, 57, 63, 90, 116, 148, 152, 167, 173
  adverse human 56
  negative 3, 152
Health issues 5, 9
  large-scale 5
Health outcomes 38, 40, 148, 184
  adverse 38, 40, 184
Health problems 67, 146, 147, 148, 150, 152, 154
  human 67
  mental 148, 150, 154
  physical 146, 147, 152, 154
  public 154
Heart 5, 6, 7, 148
  attacks 6, 148
  consultant 7
  disease 5
Heavy metals 87, 98, 99, 100, 101, 151
  accumulation of 100, 101
  arsenic 151
  bound 100
  effects of 87, 99
  increased 99
  sources of 98
Hepatitis 56, 70
Hepatotoxicity 83
Heterogeneous 102, 130
  photo-catalysts 102
  manner 130
Human carcinogen 2, 79
  strong 2
  suspected 79
Human cell transplantation 66

I

Illegal 7, 8, 39
  dumping of toxic waste 7
  urban waste dumps 39
  waste disposal sites 8
Illness 4, 19, 23, 64, 82, 123, 146, 147, 151, 152, 153, 166
  chronic 23
  life-threatening 151
  mental 146, 147, 152
  physical 146

reversible 19
temporary 23
Incidence 34, 36, 116, 126, 127, 128
  chronic disease 34
  orofacial defects 36
Incinerators 7, 17, 18, 31, 34, 35, 36, 38, 54,
    85, 173, 176, 181
  industrial waste 35
  link hazardous waste 38
  municipal 34
  single-chamber 85
Inclusive list system 20
Industrialization 5, 40, 51, 53, 73, 119, 148,
    163, 168
  increased 51, 53
  sustainable 163, 168
Industries 5, 22, 25, 43, 51, 53, 54, 57, 73, 76,
    84, 97, 98, 99, 103, 119, 147, 149, 168,
    185, 186
  allied products 33
  chemical 5
  explosive 54
  foundry 84
  historic 119
  mining 97, 98, 103
  retrofit 168
  tannery 99
Infections 5, 56, 80, 81, 172
  respiratory tract 80
  virus 56
Infertility 67, 77, 126, 127, 129, 130, 131
  couple 127
  idiopathic 131
  male 126, 127, 129, 130, 131
Ingestion 56, 67, 71, 169, 172
  plastic debris 169
Intake 85, 101
  plant 101
  reducing folic acid 85
Integrated risk information system (IRIS) 2
Intellectual development 146, 147, 152, 153,
    154, 155
  compromised 146, 147
Intelligence quotient (IQ) 41
Intensive agricultural area 28
International agency 69, 70, 76
  for cancer research (IACR) 70, 76
  research on cancer (IARC) 69
International atomic energy agency (IAEA)
    75, 150

Intervention 86, 119, 121, 134
  therapeutic 121

K

Kidney 68, 75, 79
  damage 79
  diseases 68
  loss 75
  poisoning 75

L

Land 26, 119, 121
  agricultural 26
  contaminated 119
  key 121
Landfill disposal 25, 26, 29, 30, 183
  secure 26
  technology 30
Land treatment facilities 24
Large quantity generators (LQGs) 22
Leachate 25, 26, 27, 29, 40, 88, 172, 176
  collection system 29
  detection 25
Leukemia 38, 39
  chronic lymphatic 38
Leukocyte telomere length (LTL) 132
Lifetime 3, 30, 68, 77, 115
  cancer risk 3
  expected 30
  organ's 77
Liners 25, 26, 28, 29, 30
  double composite 29
  impermeable 26
  natural clay 28
  particular 29
Liquid 20, 30, 51, 175
  washes 175
  wastes 20, 30, 51
Liver 6, 68, 71, 76, 80, 83
  cancer deaths 76
  damage 71
  detoxification 68
  dysfunction 80
  failure 6
  function 83
  toxicity 83
  tumors 71

Liver injury 83
  in hazardous waste workers 83
  tests 83
Lysosomal membrane stability 125

M

Malformations 7, 34, 37, 127
  congenital 7, 34, 127
Marine 38, 39, 169
  debris 169
  shale processor (MSP) 38, 39
Material recovery 174, 177
  Facilities 174
  process 177
Materials 43, 73, 74, 88, 102, 166, 172, 174,
    177, 180
  non-toxic 180
  organic 88
  plastic 172
  polymer 74
  raw 43, 73, 102, 166
  recyclable 166, 174
  recycle 88
  reusing 172
  secondary 174
  utilized 177
  virgin 177
Matrices 68, 74, 131
  biological 131
  environmental 68
Measuring waste management performance
    179
Mechanisms 59, 63, 68, 69, 71, 84, 85, 87, 88,
    100, 115, 122, 123, 124, 129, 130
  antioxidant defense 87
  cascading 122
  chemical synthesis 88
  exact 100
  genetic toxic 63
  metabolic detoxification 124
  physiological protective 122
Membrane chromatography 103
Mental health 146, 152, 153, 154, 155
  consequences 146
Mercury 5, 6, 7, 72, 73, 74, 80, 81, 99, 100,
    151, 173, 174, 179, 180
  containing waste 7
  pollution 99
  thermometers 173

Metabolism 67, 70, 71, 72, 153
  glucose 67
  high 153
Metabolites 3, 68, 83, 126
Metal exposure 100, 114
  heavy 100
Metallothionein 114, 115, 124, 125, 132
  biosynthesis 115
  messenger RNA 114
  proteins 114, 115
Methods 62, 176
  developing convenient management 62
  engineering 176
MicroRNAs 129, 130
  noncoding 129
Mitochondrial nitric oxide synthase 86
Mix 82, 167
  global energy 167
Modifications 83, 101, 115, 129, 130, 133
  epigenetic 130, 133
  histone 129
Multi-attribute utility analysis technique 32
Multifactorial etiology 83
Multinational company trafigura-arrived 152
Municipal solid waste (MSW) 5, 6, 34, 40, 41,
    42, 172, 173, 180
Mutations 68, 72, 90, 115, 130, 131, 148
  genetic 68, 130, 148

N

Nanomaterials 89
  green 89
  safer 89
National 4, 37, 38, 180
  disease clusters alliance (NDCA) 38
  environment agency (NEA) 180
  priority list (NPL) 4, 37
  resources defense council (NRDC) 38
Natural 31, 68, 119, 120, 165
  areas, preserving 119
  extraction impacts 165
  features 31
  resource trustees 120
  structure 68
Nausea 79
Newborns 36, 77
  low-birth weight 36
Non-gene developmental toxicities 63

Non-governmental organization (NGOs) 5, 184
Non-Hodgkin's lymphoma (NHL) 34
Non-obstructive azoospermia cases 131
Nonspecific liver disease 83
Nuclear 32, 99, 129, 130, 150
   condensation 129
   factory 32
   fission 99
   reactor 150
   remodeling 130
Nucleus 130
   human sperm's 130
   spermatozoa's 130
Nutrients 70, 174
   depleted 174
   disassemble 70

O

Occupational exposure 75
Oligozoospermia 131
   non-obstructive 131
Organs 67, 126
   sensitive 67
   sensitive system-functional 126
Oxidation 86, 97, 101
   advanced 97, 101
Oxidative stress 87, 114, 131, 132
   diseases 114
   inducing 114

P

Partnerships 149, 163, 170, 180, 182
   global 163, 170
   multi-stakeholder 170
   national voluntary 180
Passive sampling devices 37
Pathogenesis 127, 128
Pathological waste 54, 64
Pathways 70, 86, 102, 123, 146, 152, 155
   anaerobic 102
   signalling 86
Pensky-Marteus Closed-Cup Method 20
Percutaneous absorption 72
Perfluorooctanoic acid 175
Permeate 85, 88
   cytotoxic substances 85

Pesticides 38, 53, 54, 55, 57, 83, 84, 85, 99, 114, 117, 128
   organophosphate 114
Pharmaceutical waste 64
Physico-chemical methods 97
Physiological malfunctions 68, 148
Plants 28, 31, 55, 174
   coal-fired power 31
   large composting 174
   nuclear power 55
   wastewater treatment 28
Plastic waste 40, 175, 184
   amount of 40, 184
   generation 184
Pneumoconiosis 79
Poisoning 1, 71, 74
   gas 74
   rapid 71
   unsafe manner 1
Policies 17, 56, 57, 113, 134, 159, 164, 168, 176, 182, 185, 186
   legislation 56
   public prevention 134
   relevant 159
Policy 57, 165, 186
   coordination 165, 186
   framework 57
Pollutants 40, 63, 81, 90, 114, 118, 122, 128, 131, 147
   chemical 90, 147
   environmental 63, 81, 118, 122, 128, 131
   industrial 40
   inorganic 114
Pollution 167, 169, 177
   carbon 177
   mass 167
   nutrient 169
   reducing 167
   reduction marine 169
Polycyclic aromatic hydrocarbons (PAHs) 6, 8, 35, 38, 55, 81, 114, 117, 124, 129
Postremediation landscape 120
Potential daily intake (PDI) 68, 69
Pressure 30, 79, 169
   increased blood 79
   upward 30
Pressurized containers 64
Prevention 22, 26, 57, 116, 118, 121, 133, 159, 166, 169, 172
   environmental risk 121

pre-primary 116
Preventive health surveillance programs 113
Problems 18, 25, 28, 29, 33, 39, 40, 57, 58,
    59, 78, 79, 123, 146, 159, 160, 165, 184,
    185
  ever-growing 159
  growing global 146
  hazardous waste management 18
  heart 79
  kidney 79
  reported health 40
  respiratory 39, 78
  skeletal 78
Production 73, 74, 75, 79, 86, 88, 89, 99, 149,
    159, 166, 168, 172, 185, 186
  chemical 88
  crop 99
  industrial 73, 172
  residuals 176
  white blood cell 79
Products 3, 9, 17, 19, 40, 67, 88, 89, 166, 172,
    173, 174, 175, 176, 179, 180, 184
  chemical 88, 89
  cleaning 17
  commercial chemical 19
  consumer 9
  eco-effective 176
  nontoxic 172
  packaging 40, 184
  toxic 88
Progenitors 72, 90
Progressive disease 65
Prostate 9, 67, 116
  diseases 67
  glands 67
Protection 90, 119, 131, 153, 161
  antioxidant 131
  cultural 119
  environmental 90
  social 161
Protective 58, 62, 85
  clothing 85
  measure 58
  proceedings 62
Protein products 84
  tumor gene 84
  tumor-related 84
Proteins 72, 82, 83, 84, 114, 115, 129, 130
  alternative 82
  arginine-rich 129

heat shock 115
human 82
inducible 114
key 130
plasma 72
stress-damaged 115
tumor-related 84
Psychological distress 127
Public awareness 6, 9, 182
  high 182
  low 6

Q

Quality 1, 3, 7, 10, 98, 146, 148, 152, 155,
    167
  crop 98
  perceived 146, 152, 155
Quality control techniques 181
Quality education 149, 161, 163, 167
  equitable 163, 167

R

Radiation 150, 153, 155
  contamination accidents 153
  disaster 150
  incident 155
  poisoning 150
  sickness 150
Radioactive 99, 150
  debris 150
  isotopes 99
Reactive 21, 85
  chemical entities 85
  wastes 21
Reactivity 19, 20, 21, 86, 148
  high chemical 86
Rechargeable batteries 79
Recycle 179
  glass 179
  metal 179
  paper 179
  plastic 179
Recycling 5, 9, 17, 23, 24, 25, 41, 88, 166,
    168, 174, 180, 182
  informal 5
  proper 180
  waste 17

Reduction of hazardous wastes generation 42
Regulations 10, 19, 21, 33, 41, 59, 115, 129,
    130, 185
  existing 10
  gene expression 129
  posttranscriptional 130
Renewable energy 167
Reproductive system 77, 78, 118, 126, 128,
    131, 134
  male 118, 126, 131
  toxins 77
  tract, male 128
Risk assessment 2, 62
  process 2
  sounds 62

S

Sectors 6, 28, 55, 123, 149, 162, 168, 182
  commercial 55
  intensive industrial 28
  private 6, 162, 182
Semen quality 118, 128, 133
  human 118, 128
  linked 133
Short-chain chlorinated paraffins (SCCPs) 175
Skin 8, 34, 63, 66, 67, 71, 72, 75, 78, 80, 101,
    152
  absorption 67
  burns 152
  infections 75
  irritation 80
  rashes 78
Sludges 19, 97, 101, 147, 151, 153
  activated 97, 101
  iron tailings 151
  iron waste 151
  toxic 153
Small quantity generators (SQGs) 22
Smoke 39
  tobacco 76
Smoking 70, 76, 84, 99, 100, 101
  cigarette 76, 101
  tobacco 70
Socio-economic 57, 85
  development 57
  repercussions 85
Solids 19, 20, 57, 147, 171
  discharging 171

Solid waste 5, 10, 19, 34, 51, 52, 54, 88, 99,
    146, 165, 172, 173, 175, 178
  amendments 54
  leakage 175
  toxic 10
  urban 178
Soluble threshold limit concentrations
    (STLCs) 21
Sorption methods 104
Spatial 38, 39, 126
  movements 39
  stable 38
Sperm 77, 82, 85, 129, 130, 131
  alterations 131
  aneuploidies 129
  mature 130
Spermatic DNA fragmentation 132
Spermatocytes, primary 77
Spermatogenesis 130, 131
Spermatogonia 77, 127
Spermatozoa 77, 127, 129, 130, 131
  mature 130
Sperm DNA 127, 129, 132
  damage 127, 132
  fragmentation 132
  methylation 129
Spermiogenesis 130
Sperm telomere length (STL) 132
Strategy 17, 32, 33, 42, 87, 90, 126, 159, 160,
    166, 170, 173, 179, 182, 186
  effective primary prevention 126
  harmonic 87
  national resource management 186
  national waste 166
  risk substitution 32, 33
Stress 29, 81, 100, 115, 122, 124, 125, 126,
    129, 130, 131, 133, 155
  body's 125
  chemical 122
  epigenetic 131
  excess 100
  exogenous 133
  external 130
  physical 126
  physiological 81
Structural modification 116, 117
Substances 8, 62, 65, 66, 67, 68, 69, 70, 71,
    72, 73, 74, 78, 79, 80, 81, 84
  cancer-inducing 84
  cytotoxic 85

dangerous 62
destructive 67
fat-soluble 72
non-radiological 72
non-threshold 69
noxious 62
safe 71
synthetic 65
venomous 66
Susceptibility 62, 114, 117, 124, 126, 128,
    131
  elevated 114
  genetic 126
  increased 114
Sustainability 120, 164, 169, 171, 172, 174,
    185
  environmental 164
  materials management 185
  ocean 169
Sustainable development 1, 42, 121, 159, 162,
    164, 165, 166, 167, 168, 169, 170, 171
  goals (SDGs) 1, 42, 159, 162, 164, 165,
    166, 167, 168, 169, 170, 171
  practices 121
Sustainable 146, 148, 150, 173, 186
  consumption behavior 186
  economic development 148, 150
  natural cycles 173
  societal development 146
Sustainable waste management 1, 40, 168,
    170, 171, 172, 185
  efforts 171
  strategies 1
Synthesis 87, 89, 97, 98, 103, 114
  chemical 97, 98, 103
  nanomaterial 89
Synthetic 67, 77
  estrogen 67
  steroids 77
System 5, 19, 20, 33, 35, 41, 42, 85, 114, 122,
    124, 147, 169, 174, 175, 176, 178, 179,
    183
  activating multi-enzymatic 122
  advanced geographical information 35
  anaerobic digestion 176
  cellular antioxidant 114
  connective tissue 147
  digestive 5
  environmental 169
  hormone 85

immune 124
organized waste disposal 33

T

Target hazard quotients (THQ) 100
Techniques 9, 23, 89, 104
  biomarker epidemiology 9
  explored treatment 104
  green nanomaterial generation 89
  pollution prevention 23
Technologies 7, 9, 88, 102, 103, 148, 168
  biogas 88
  communications 168
  evolved 148
  hazardous waste treatment 25
  membrane 103
  modern 9
  multiple 24
  remedial 102
  sound hazardous waste treatment 7
Thyroid-stimulating hormone 4
Tissues 8, 56, 74, 122, 126, 169
  adipose 74
  fat-rich 8
  malignant lung 169
Tools 3, 7, 37, 89, 116, 176, 177
  alternative performance assessment 176
  biological/biochemical 7
  life cycle assessment 177
Topography 27
Total threshold limit concentrations (TTLCs)
    21
Toxicants 73, 81, 82, 114, 127
  environmental 114
Toxic inorganic impurities 101
Toxicity 69, 77
  evolutionary 77
  gene 69
  reproductive 77
Toxicological monitoring 123
Toxic
  release inventory (TRI) 38
  Substance Control Act 54
  substances 21, 22, 63, 64, 66, 67, 70, 72,
    74, 83, 85, 152, 155, 173, 176
Toxic waste(s) 3, 7, 21, 55, 64, 85, 88, 146,
    147, 152
  and solid waste 146
  producers 88

Toxic wastes dumping 7, 152, 153
  massive 152
Transport 27
  airborne 27
  railways 27
Trauma 81, 152, 153
  psychological 152
Treatment 18, 23, 24, 25, 26, 57, 58, 64, 87,
    90, 101, 118, 179, 180
  biological 24
  chemical 101
  dietary 118
  special end-of-life 179
Tumors 3, 63, 67, 69, 73, 84, 115, 116
  colon 84
  hormone-sensitive 116
  pre-cancerous 67
  pre-tumor colonic 84

U

Ulcers 79
United nations environmental program
  (UNEP) 7, 19, 28
UN statistical commission (UNSC) 162

V

Virgin material inputs 177
Volatile 55, 132
  endogenous 132
  and explosive waste construction 55
Vomiting 79, 152

W

Waste 1, 4, 6, 7, 17, 19, 20, 21, 22, 24, 25, 30,
    34, 39, 40, 42, 51, 54, 55, 57, 64, 75, 76,
    85, 88, 99, 102, 147, 151, 166, 172, 173,
    178, 179, 184
  biomedical 64
  chemical 4, 64, 173
  corrosive 20, 21
  crisis 7
  cytotoxic 85
  decomposing 39
  dispose 184
  drinking 40
  dry 178, 179

electronic 99
fruit 102
generic 19
green 178, 179
healthcare 6
heavy metal 99
hospital 172
household 34
iron 151
manufacturing 57
petroleum company 147
radioactive 54, 64, 75, 76
safe 166
toxic coal fly ash 151
Waste disposal 5, 9, 21, 30, 51, 146, 150
  municipal solid 5
  operations 30
  toxic 21, 150
Waste electric and electronic equipment
  (WEEE) 179, 180, 181, 182
Waste generation 39, 40, 98, 166, 178, 184,
  185
  elevated 40, 184
  solid 98
Waste management 5, 6, 9, 10, 17, 39, 40, 43,
    56, 57, 78, 159, 165, 166, 167, 168, 170,
    171, 172, 173, 182, 183, 184, 185
  changes in 159, 184
  facilities 56, 78
  illegal 170, 183
  improper 5, 168, 171
  inadequate toxic 10
  municipal 5
  proper 166, 170, 184
  real sustainable 184
  safe 17
  solid 183
  strategy 43, 165, 182
  sustainable hazardous 57
Waste management systems 41, 176, 177,
    178, 183, 184, 185
  proper 183
  real integrated 173
  sustainable 184
Waste sites 3, 41, 78, 83, 90, 147, 153
  cleaning hazardous 90
  contaminated hazardous 41
  contaminated toxic 153
  managed hazardous 147
Water 40, 172, 177, 179

pollution 172
quality 40
resources 99
saving 177, 179
scarcity 167
Workers 2, 4, 56, 64, 65, 71, 78, 84, 119, 132,
        151, 153, 168
and community residents 153
exposed 84
health care 56
on-site 2
on-site remediation 119
waste site remediation 78
World commission on environment and
        development (WCED) 160

Y

Years of life lost (YoLL) 36

Z

Zero waste 41, 165, 173, 174, 176, 177, 178,
        179, 185
concept 173, 185
development 174
index (ZWI) 176, 177, 178, 179
infrastructure 173
initiatives 41, 165, 173
Zero waste strategy 41, 170, 174, 175, 176,
        177, 179, 181, 182, 183, 186
adopted 183
Zero waste system 42, 173, 179, 182
adopted 42
global 173

www.ingramcontent.com/pod-product-compliance
Lightning Source LLC
Chambersburg PA
CBHW050839220326

41598CB00006B/398